- reduce pollution
  increase η (we
→ proven nure impr
  reduce demand
  96% from steam reformation

Hydrogen and Fuel Cells:
A Comprehensive Guide

Hydrogen and Fuel Cells:
A Comprehensive Guide

Rebecca L. Busby

This book represents an overview of hydrogen as a fuel source, as well as uses for fuel cells and the potential for a hydrogen-based economy in the United States. As such, any and all opinions concerning equipment, manufacturers, technology, and utility and non-utility company strategies or services are the personal opinions of the author and may not necessarily reflect the opinions, positions, or claims made by any manufacturers or utilities herein mentioned.

Copyright © 2005
PennWell Corporation
1421 South Sheridan Road
Tulsa, Oklahoma 74112

1-800-752-9764
sales@pennwell.com
www.pennwell.com
www.pennwell-store.com

Managing Editor: Stephen Hill
Cover Design: Charles Thomas
Book Design: Wes Rowell

Library of Congress Cataloging-in-Publication Data

Busby, Rebecca L.
  Hydrogen and fuel cells : a comprehensive guide / by Rebecca Busby.--
1st American ed.
       p. cm.
  Includes bibliographical references.

  ISBN 1-59370-041-5

  1. Hydrogen as fuel. 2. Fuel cells. I. Title.
  TP359.H8B87 2005
  621.31'2429--dc22

                                                                    2004029903

Printed in the United States of America
1 2 3 4 5   09 08 07 06 05

Contents

List of Figures

List of Tables

Acknowledgments

In writing this book, I was inspired by my memories of working with several pioneers in hydrogen and fuel cell research during the 1970s and 1980s, including Henry Linden, Derek Gregory, Bill Escher, Jon Pangborn, and Root Woods. Those were the good old days, when youth and optimism made everything seem possible.

This book also relies on the hard work of many other authors, including Peter Hoffmann, Vijay Vaitheeswaran, Joan Ogden and Daniel Sperling, Jeremy Rifkin, Amory Lovins, and countless other energy analysts and engineers. All of these people have laid the foundation for global efforts to incorporate hydrogen in our future energy systems.

I thank my family and friends for encouraging me along the way while enduring a long period of neglect, and I dedicate this book to my husband Dan, who has read all the drafts. As an engineer with a talent for critical thinking, he's been a diligent nitpicker, as well as my computer troubleshooter and cheerleader.

What's all the Excitement About?

Imagine an unlimited supply of pollution-free power. That's the ultimate reason for excitement about hydrogen. Right now, it's only a theoretical promise. How we might actually get there is the subject of this book.

Hydrogen's main attraction to energy engineers is that it can be produced from water or from any hydrocarbon fuel, or even from energy sources of the future that we can't envision yet. Like electricity, hydrogen carries energy that is produced from other sources and delivers it in the form of power and heat. Compared to electricity, however, hydrogen is easier to store—a vital distinction.

The feature of hydrogen that excites environmental scientists is that it creates no harmful emissions when used in an electrochemical fuel cell to generate power. The only by-product is water. In the future, all of our energy needs could be met by hydrogen that is produced from renewable sources like wind power and solar energy.

The transition toward a hydrogen economy is already under way. Over the past century or so, we've switched from wood fuel to coal, then to oil, and now to natural gas, which is gradually taking oil's place as our most popular fossil fuel. This shift reflects a slow reduction in the amount of carbon contained in the fuel and an increase in its hydrogen content. The ultimate next step is to eliminate carbon altogether and use pure hydrogen.

Hydrogen fuel cells are already being tested in vehicles and used commercially in micropower plants, which generate electricity where it's needed instead of transmitting it over long distances. In a future hydrogen economy, fuel cells would power most of our vehicles, and micropower plants would proliferate, complementing our supply of electricity from the utility grid by providing extra power during periods of peak demand.

In the ultimate vision of a hydrogen economy, we might be able to fill up our cars from a hydrogen dispenser at home or plug them into our electrical system to generate power for sale to the utility. Eventually, everyone might be connected to an energy internet or hydrogen energy web, and the whole system could be managed by virtual utilities.

None of this would be conceivable, however, without powerful forces at work driving us toward a global hydrogen economy. Nor would anything like this happen without the blessing of the automotive industry, oil companies, and utilities, as well as financial support from governments worldwide.

What's exciting is that these factors are all beginning to fall into place. Industrial nations are concerned enough about global warming to begin regulating emissions of carbon dioxide. Unremitting turmoil in the Middle East is compelling us to reconsider how and where we get our energy. Automakers have joined forces with big oil to test hydrogen fuel cell vehicles and hydrogen fueling stations around the world. Utility customers are demanding higher quality, uninterruptible power. And governments are supporting hydrogen research at unprecedented levels.

Of course, many barriers would need to be overcome to realize hydrogen's promise. Technology development would have to reduce the costs of manufacturing fuel cells, building a fueling infrastructure, capturing carbon dioxide emissions, and producing, storing, and transporting hydrogen. For example, the cost of delivering hydrogen to dispersed customers today is extremely high—up to five times more expensive than producing the hydrogen. Many regulatory and institutional barriers also would have to be alleviated.

But many of the world's best minds think it can be done—and that it must be done, for the sake of future generations. Imagine a world where our children and grandchildren will breathe cleaner air, where the poorest village can make electricity from the sun, where no one needs to fight over energy supplies. Now that's exciting.

Chapters 1 and 2 explain the potential of a future hydrogen economy, the forces driving a global transition toward hydrogen, and the barriers that could derail such a transition. Hydrogen's properties and applications are covered in chapters 3 and 4. The production, transport, storage, and safety of hydrogen are discussed in chapters 5, 6, and 7.

Concept and Vision

Hydrogen is not a source of energy like oil or natural gas—instead, production of hydrogen consumes energy. That's because hydrogen atoms are almost always bonded with other elements into compounds like water or hydrocarbon fuels, which require energy to break apart.

However, because hydrogen is found in so many different fuels and compounds, it can be produced from any energy source—and it can be obtained from one of our planet's most common substances, water.

The concept of a hydrogen economy envisions a future where all of our energy needs will be met by hydrogen that is produced from renewable sources like wind power and solar energy. Our economy would be based on an unlimited supply of pollution-free power. This vision is not just around the corner, but a transition is gathering steam, and hydrogen certainly has a part to play in the world's energy mix within the foreseeable future.

Hydrogen as an Energy Carrier

Rather than being an energy source, hydrogen is usually considered an energy carrier—a way to transport and store energy that is produced from other sources. Similarly, electrons serve as an energy carrier today in the form of electric power.

Like electricity, hydrogen does not exist freely in a usable form, but it can be made using natural gas or other primary energy sources. The main difference is the end product, electrons or hydrogen molecules. Compared to electricity, however, hydrogen is easier to store.

As an energy carrier in the future, hydrogen could free us from transporting fossil fuels—piping natural gas, trucking gasoline, and importing oil. Hydrogen could bring the same fuel energy to us, but in a different useful form. In a hydrogen economy, all we are really doing is shifting the delivery method from hydrocarbons toward hydrogen.[1]

Production From Any Source

One of hydrogen's advantages is that it can serve as a basic "unit of currency" in the energy marketplace. Again, this is similar to electricity. The use of electrons as a common denominator throughout our energy systems greatly simplifies our lives. The energy services we receive do not depend on how those electrons are produced. What people get are lights and air-conditioning on demand, not energy *per se*.

Hydrogen would excel in the same way. Both hydrogen and electric power can be made from any fossil fuel or from any renewable resource, such as wind and solar energy. "The beauty of the hydrogen model is that it is not wedded to any specific primary energy source or technology," said Vijay Vaitheeswaran in his book, *Power to the People*.[2]

Hydrogen's flexibility of production would also allow it to be phased into our existing energy systems in whatever way is least disruptive to any particular locale or economy. For example, in the sunny Israeli desert, hydrogen might be produced directly from solar power, while in coal-rich spots of the United States, gasified coal might be used to make hydrogen, if clean processes can be developed.[3]

And if new sources of energy are discovered and developed in the future, they also could be used to make hydrogen—without affecting delivery of energy services to the customer. We wouldn't have to change the way we obtain or use our energy, since it would be brought to us in the same way. Hydrogen could complement electricity as an alternative energy delivery service.

Hydrogen Complements Electricity

Together, hydrogen and electricity can satisfy most of our future energy needs and, in fact, would compete with each other to do so. In the foreseeable time frame, hydrogen would coexist as another form of energy alongside electricity. Hydrogen would win out in applications where it has an edge over electricity, and vice versa.

Hydrogen and electricity are closely related. You can use electricity to produce hydrogen, and you can use hydrogen to produce electricity (see chapter 3). This means that if hydrogen and electricity were both in widespread use, they could be substituted for each other fairly easily. This flexibility would be valuable in terms of getting the most use out of our existing equipment and infrastructure, and our economy's energy systems could become even more versatile than they are today.

In the more distant future, an all-electric economy might be the only alternative to hydrogen. But enormous strides would be needed to develop better technologies for storing electricity, and power generation would have to become much cleaner and more efficient. Instead, many engineers think that hydrogen and electricity will complement each other quite nicely in the future—renewable sources would generate electricity, and hydrogen would store and deliver it.

The U.S. Department of Energy, in its 2004 hydrogen plan, was careful to include electricity in its description of a future hydrogen economy: "Hydrogen will become America's 'clean energy choice,' joining electricity as a primary energy carrier and providing the foundation for a globally sustainable energy system."[4]

Taking the C Out of HCs

One of the most compelling reasons to promote a hydrogen economy is to improve our planet's environment for future generations. Hydrocarbon fossil fuels create carbon dioxide when they're burned, as well as many other air pollutants and toxic chemicals (see chapters 2 and 3).

Hydrogen represents the end point of humanity's trend toward using less and less carbon—a trend that has persisted throughout our history of consuming modern-day fossil fuels and for thousands of years before that, if you count our ancestors burning wood, peat, and dried-out animal waste.

These primitive fuels were succeeded by coal as our primary energy source, and since then coal has shared its place with oil. Progressively, these fuels contain less carbon and more hydrogen, in terms of atoms per molecule. Wood's carbon-to-hydrogen ratio is roughly 10:1, while coal ranges from 2:1 to 1:1 (equal amounts of carbon and hydrogen). Oil slides farther down the carbon scale with a 1:2 ratio, or an average of 1 carbon atom for every 2 hydrogen atoms.

Today natural gas is gaining on oil's market share and is perceived as its successor in many applications. In America and some developing countries, new power generating capacity has shifted toward natural gas, which in its liquid form is now shipped worldwide. And as you can guess, natural gas has even less carbon and more hydrogen than oil (a 1:4 ratio).

So with each step toward the present, we have burned less and less carbon (Fig. 1–1). Globally, this trend has reduced carbon emissions per unit of primary energy consumed by about 0.3% annually for the past 140 years.[5] As Amory Lovins of the Rocky Mountain Institute puts it, "Today over two-thirds of the fossil-fuel atoms we burn are hydrogen, not carbon. The next step eliminates both the burning and the carbon."[6]

HISTORIC WORLD PRIMARY ENERGY SUBSTITUTION

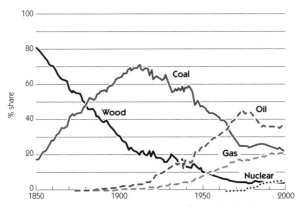

Source: Dr. Nebosja Nakicenovic, International Institute for Applied Systems Analysis, Laxenburg, Austria. Private communication, Aug. 20, 2003.

FIG. 1–1. TREND TOWARD LOWER CARBON FUELS. SINCE WOOD FIRST DOMINATED WORLD ENERGY USE, SUBSEQUENT FUELS HAVE MOVED PROGRESSIVELY TOWARD LOWER CARBON CONTENTS. NATURAL GAS IS GRADUALLY OVERTAKING OIL AS THE MOST POPULAR PRIMARY ENERGY SOURCE. ARTWORK COURTESY OF *Oil & Gas Journal*.

Three factors—comfort, convenience, and cleanliness—account for the shift away from carbon, says Vaitheeswaran. "These are the best reasons to think that the world will one day reach the hydrogen era." By 2050, natural gas and hydrogen could replace oil and coal and eventually command 75% of the global energy market—"the same as King Coal in his heyday."[7]

In today's scientific jargon, this process is called decarbonization, a term coined in 1989, which also refers to the artificial extraction of carbon from fuels. A decarbonized fossil fuel is, simply, hydrogen. However, to produce hydrogen from today's fuels, we would need to store the carbon somewhere safe, which is called sequestration (see chapters 2 and 5).

Storing Electricity From Renewable Sources

Some energy technologies become much more practical when they in-corporate storage—for example, solar and wind power, which produce varying amounts of electricity depending on the weather and the time of day. Hydro-gen could serve as an energy storage bank for these intermittent renewable sources—a valuable function, because electricity is difficult to store.

Whenever the wind is blowing or the sun is shining, the power that's gen-erated by a wind turbine or photovoltaic panel can be used to make hydro-gen. When the utility grid is running at maximum capacity and needs more power, the hydrogen can be reconverted into electricity and sold to the grid, perhaps at premium prices. Hydrogen would serve the purpose of smoothing out the daily and seasonal fluctuations of solar and wind power.[8]

Although renewable sources still provide only a small fraction of our electricity, wind power has been the world's fastest growing energy source for the past several years, surpassing increases in other renewables and in conventional fuels as well. Globally, wind energy capacity grew by 35% per year on average during 1999–2003 and in the United States by 28% annu-ally. Eventually, hydrogen could be used to store this vast, global resource (Fig. 1–2).

FIG. 1–2. NEG MICON WIND TURBINE. WIND POWER IS THE WORLD'S FASTEST
GROWING ENERGY RESOURCE, ESPECIALLY IN EUROPE, WHERE THESE WIND
TURBINES WERE INSTALLED IN TEBBESTRUP, DENMARK. PHOTO COURTESY OF
NEG MICON A/S.

Hydrogen itself, once it has been produced from renewable energy or
other sources, can be stored in a variety of ways (see chapter 6). Traditionally,
hydrogen gas has been compressed or liquefied for storage in tanks. It can
also be used to produce other storable liquid fuels such as methanol and
ammonia. In technologies now being developed, hydrogen can be attached
to certain metals to form hydrides, and scientists are inventing ways to store
hydrogen on microscopic carbon fibers.

Handling and Producing Hydrogen

Many people might worry about the safety of hydrogen energy
systems—driving around with hydrogen in their fuel tanks or relaxing in
their homes with a hydrogen-fueled power generator in the back yard or
basement. Notorious disasters have exaggerated the public's perception of
hydrogen's dangers for decades.

Hydrogen is relatively safe compared to gasoline, propane, and other common fuels. As Vaitheeswaran points out, "If gasoline had to win approval by environmental authorities as a new fuel today, it could well be rejected as too dangerous."[9]

Because hydrogen is so lightweight, it would disperse upward if it leaked, instead of spreading into a dangerous pool that could be ignited. By the same token, hydrogen can't spill into the ground and contaminate the water supply as oil can, nor could it form a huge oil slick on the ocean.

In a future hydrogen economy, we might eventually store and transport hydrogen in ways very similar to how we handle natural gas and propane today (see chapter 6). But research is needed on potential leakage in enclosed structures, and codes and standards would have to be further developed in concert with technological improvements.

The costs of producing and transporting hydrogen also need to be addressed. For example, the cost of delivering hydrogen to dispersed customers today is extremely high—up to five times more expensive than producing the hydrogen. Also, the process of making hydrogen from compounds like water and natural gas consumes more energy that we get when we subsequently use the hydrogen as a fuel. Cost barriers are discussed in chapter 2, and production and transportation research and economics are described in chapters 5 and 7.

Power On Demand From Hydrogen Fuel Cells

Fuel cells began to capture the public's imagination—and venture capitalists' dollars—in the 1990s. But their basic technology is more than 160 years old, which is 40 years older than our everyday internal-combustion engine. Space exploration was the first practical application for fuel cells, beginning in the 1960s (see chapter 3).

Fuel cells could play a critical role in the hydrogen economy by producing power on demand—the end point of the hydrogen delivery service. Today, fuel cells are recognized as a vanguard technology that could launch hydrogen into the world's energy mix.[10] Hydrogen can also be used in engines and turbines to generate power.

You can think of a fuel cell as a type of battery that produces electricity. But conventional batteries generate power from stored chemicals, which need to be replaced. In contrast, chemical energy in the form of a fuel is fed into the cell, so it can generate electricity as long as it's getting fuel and oxygen (Fig. 1–3). Hydrogen is the fuel cell's ideal fuel.

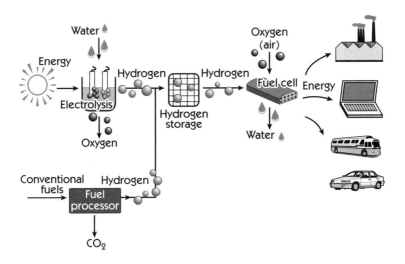

FIG. 1–3. HYDROGEN ECONOMY CONCEPT. EVENTUALLY, RENEWABLE SOURCES WOULD PROVIDE HYDROGEN FOR FUEL CELLS, BUT NATURAL GAS AND OTHER CONVENTIONAL FUELS CAN BE USED DURING THE TRANSITION. ARTWORK COURTESY OF SHELL.

Fuel cells work by combining hydrogen with oxygen from the air in an electrochemical process that is clean and flameless. Nothing is produced but electric power and pure, distilled water, along with useful heat (see chapter 3).

Today, fuel cells are emerging in the energy marketplace, not only in experimental vehicles but also in small power plants serving groups of homes and buildings and in larger sizes powering factories and other industrial facilities.

Today's fuel cells generate power much more efficiently than a car engine running on gasoline—roughly 2½ times more, according to most estimates. The U.S. electric power network converts only 33% to 35% of fuel energy into electricity on average and wastes most of the heat released during power generation.

In contrast, fuel cells capture 40% to 65% of hydrogen's energy and convert it to electricity.[11] However, the service life of today's fuel cells is much too short, and fuel cell costs are 10 to 20 times too high for mass-market success. Fuel cell efficiencies and costs are discussed in more detail in chapters 2 and 3.

Fuel cell markets

In a future hydrogen economy, we would use fuel cells in three main ways:

1. Transportation, which includes fuel cell-powered passenger cars, pickup trucks, minivans, commercial vehicles, transit buses, locomotives, ships, and planes—just about any mode of moving, shipping, or carrying that you can imagine

2. Stationary, referring to fuel cell plants generating electricity from a fixed location, which can range in power output from a few kilowatts in units designed for homeowners up to the multi-megawatt range for utility power stations

3. Portable, which refers to miniature fuel cells powering laptop computers, cellular phones, and other hand-held devices and small appliances.

In some concepts of a hydrogen future, these handy categories become a little fuzzier—for example, a fuel cell car might eventually be plugged into an electric outlet at home to generate power for sale to the utility. That would be a transportation fuel cell being used part-time as a stationary micropower plant.

These three applications for fuel cells are in varying stages of technology development and market maturity. Right now, the world's major automakers are busy showing off experimental fuel cell cars—and granted, they do look hot. But stationary fuel cell power is the most mature application, even though costs remain prohibitively high, and most analysts predict that fuel cells will enter mass markets for the first time in portable devices like laptops.

Fuel cell types

The technology of the fuel cell itself also falls into a handful of categories based on the type of electrolyte it uses (electrolytes are liquids or solids that channel charged particles inside the cell). These types include phosphoric acid, molten carbonate, and solid oxide, as well as polymer electrolyte membrane (or proton exchange membrane), which is dubbed PEM (see chapter 3).

Each type of fuel cell is suitable to particular applications—for example, vehicles use mostly PEM fuel cells, which are lightweight and easy to start up, while industrial facilities might select solid oxide fuel cells, which release large amounts of high-temperature heat that can be recovered and used.

One of the advantages of fuel cells is their modular design—it's relatively easy to connect several fuel cells together if more power is needed, rather than installing a single, larger system. Fuel cells can be installed just about anywhere, in a wide range of sizes. This flexibility makes them ideal candidates for the world's growing market for small power plants.

Micropower Makes Sense

Micropower is the latest term for a sensible idea: generating electricity where you need it, instead of getting it from far away. This alluring concept is also called distributed generation, a clunkier name but one that appears widespread in the literature. Micropower is an elegant solution to the inconvenience of importing our electricity along transmission lines that clutter the sky and lose power along the way.

Any small power plant that is located physically close to its customer is a distributed generator (small usually meaning less than 20 megawatts of output). These plants can incorporate fuel cells as their power source, as well as microturbines, engines, and larger turbines. Usually these generators are considered stationary plants, though some entrepreneurs are renting large engines on trailers that can be moved every few months or years.

Many micropower systems recover the heat that is given off by the fuel cell or other power source. By using this heat for practical purposes such as water heating, the whole system greatly increases its overall energy efficiency. Such systems are called cogeneration or combined heat and power (CHP). Cogenerating engine and turbine systems using fossil fuels can reach 70% to 80% efficiencies. When a fuel cell's heat is recovered and used, systems can approach efficiencies of 85%.[12]

Research has been under way for many years on fuel cells for micropower applications in homes and businesses. The first residential fuel cell in the United States, a 7-kilowatt unit, began operating in 1998 at a ranch-style brick home near Albany, NY (Fig. 1–4), and thousands more have been in-stalled since then, analysts estimate. In addition, about 600 larger fuel cells (more than 10 kilowatts of output) are providing electricity to commercial and industrial buildings worldwide (see chapter 4).

FIG. 1–4. RESIDENTIAL FUEL CELL. A 7-KILOWATT HYDROGEN FUEL CELL BEGAN OPERATING IN NEW YORK STATE IN 1998, THE NATION'S FIRST SUCH INSTALLATION. PHOTO COURTESY OF PLUG POWER.

Microgenerators

Fuel cells have been anointed as the best bet for future micropower systems in the hydrogen economy, but they aren't the only new kind of micropower source that could use hydrogen. Miniature turbines are being developed that could eventually operate on hydrogen fuel. Unlike their larger counterparts, these microturbines have very few moving parts and can do without lubrication, making them cheaper to run and maintain.

Also, renewable resources like wind turbines and solar energy are sources of micropower. As we have seen, however, they need a way to store their electrical output in order to become more practical. In this case, hydrogen would be the product of the renewable generator rather than its fuel.

Other micropower technologies being developed include engines and turbines modified to run on hydrogen or hydrogen/natural gas blends. Some hydrogen vehicles use internal-combustion engines (see chapter 4).

Receptive utilities at last?

Electric utilities traditionally have resisted the encroachment of micropower on their turf, but several factors have recently combined to stage a turnaround in utility attitudes—soaring electricity consumption during peak hours, gradual deregulation of electric industries worldwide, market pressure from eager competitors, and good old-fashioned customer demand for more reliable, higher quality power.

Also, the cost of building new transmission lines and electricity distribution systems has escalated, along with public opposition to such projects. This has frustrated utilities that desperately need more capacity to deliver power to their customers. And new technologies are affording utilities better control over unpredictable amounts of electricity streaming in and out of their grid from external sources like fuel cells and other micropower units. These factors are discussed in more detail in chapter 2.

Micropower and large natural gas-fueled turbines are expected to provide nearly 62% of new generating capacity in the United States through 2025.[13] According to the Electric Power Research Institute's former president Kurt Yeager, "Our society is changing more broadly and more rapidly than at any time since Edison's day. The current power infrastructure is as incompatible with the future as horse trails were to automobiles."[14]

The renaissance in micropower is very promising to many industry observers. Vijay Vaitheeswaran says, "If micropower really takes off, then there is every reason for optimism about our planet's future. Let the revolution roll!"[15]

Fuel Cell Cars Unveiled

Most of us have heard of someone who's gone out and bought one of those new hybrid cars, a Toyota Prius® or Honda Civic, which have enjoyed surprising popularity in the automotive marketplace. These cars come equipped with a normal gasoline engine, as well as a quiet electric motor and battery system that captures energy that would otherwise be wasted and greatly increases the car's fuel efficiency. Many new hybrid vehicles, including SUVs and pickup trucks, are entering the market (see chapter 4).

What many people don't know is that the next generation of energy-efficient cars and trucks could use hydrogen-powered fuel cells. Development of these vehicles would not have been possible without years of financial support and collaboration among automakers, oil companies, and politicians, spurred on by environmental advocates.

Right now, we still don't have a practical, mass-market alternative to transportation based on oil products. But over the long run, hydrogen fuel cells could help us break oil's grip on the transportation sectors of the world's industrial nations. With fuel cell-powered cars, we could retain our mobility but free ourselves of the air pollution that trails behind us as we drive gasoline and diesel vehicles.

Most of the world's major automakers have jumped onboard the hydrogen fuel cell bandwagon (see chapter 4). For example, Ford Motor Co. has six prototype fuel cell vehicles in development. The company's chairman Bill Ford is on record as saying, "I believe fuel cells will finally end the 100-year reign of the internal-combustion engine."

As another example, General Motors Corp. tested three fuel cell cars in the Tokyo area during 2003–04. The HydroGen3 vehicles, being operated by Federal Express Corp. (Fig. 1–5), can travel 250 miles (400 kilometers, or km) at speeds up to 100 miles per hour (160 km/h). GM has also developed concept cars that combine hydrogen fuel cells with drive-by-wire technology.[16]

FIG. 1–5. GM HYDROGEN3. THIS FUEL CELL CAR HAS BEEN TESTED BY
FEDEX IN THE TOKYO AREA. PHOTO COURTESY OF GM.

Some oil companies—most notably, the Royal Dutch/Shell Group,
BP plc, and ChevronTexaco Corp.—are also promoting fuel cell cars and supporting hydrogen energy by demonstrating hydrogen fueling technologies.
For example, BP has built the fueling infrastructure for several fuel cell bus
fleets worldwide (Fig. 1–6).[17]

FIG. 1–6. BP HYDROGEN FUELING STATION. THIS HYDROGEN STATION SERVES
THREE FUEL CELL BUSES IN BARCELONA, SPAIN, AS PART OF A EUROPEAN
DEMONSTRATION PROJECT. PHOTO COURTESY OF BP.

Oil companies are no longer just purveyors of petroleum, but have become integrated energy companies willing to supply any fuel that their customers want. Chapters 2 and 4 discuss the participation of automakers, the oil industry, and other potential players in a hydrogen economy.

The market entry and success of fuel cell vehicles will still depend on government policies, technology and cost improvements, and the outcome of competition from other options. Realistically, automakers would still need to ramp up their investments substantially to mass-produce fuel cell vehicles and launch them into the marketplace.

Profiting From Your Engine

Why stop with powering our cars? Essentially, the fuel cell car's engine is actually a tiny power plant that runs on hydrogen. In the future when you pull into your garage, you might be able to plug your car into an electric outlet, use its hydrogen fuel to generate electricity, and sell the power to the utility at a profit.

This idea, popularized by Amory Lovins of the Rocky Mountain Institute, has intrigued engineers and spurred development of technologies that would help make the concept practical. Lovins estimates that at 20 kilowatts of power output per car, a fleet of 200 million hydrogen fuel cell vehicles could have four times the generating capacity of the U.S. national utility grid.[18]

As another way of looking at it, the power contained under the hoods of America's existing cars is ten times that of the country's nuclear, coal, and natural gas power plants combined, says Vijay Vaitheeswaran. "Ford Motor Company alone could add more juice to America's power grid than all of America's conventional power utilities put together."[19]

If homeowners could sell their cars' power to the utility, argues Amory Lovins, financial subsidies might not be needed for further development of fuel cells. Your fuel cell car, while parked during the day at work—especially at times of peak consumption on hot afternoons—could be transmitting electricity to the utility grid, reaping premium prices for its owner.

In a future hydrogen economy, public fueling stations might be supplemented by smaller hydrogen production machines, and you could operate a unit at home to fill up your car's hydrogen tank and serve your home with heat and power as well. This concept is being developed by Honda R&D Co., Ltd., and Plug Power (see chapter 4).

Their experimental Home Energy Station, located at Honda's California research facility, produces hydrogen from natural gas for use in fuel cell vehicles while supplying electricity and hot water to the home. Throughout the day, the system produces enough hydrogen to fill up the tank of a Honda FCX hydrogen fuel cell vehicle in just a few minutes. The product could enter the market around 2015–2020.

A Vision of an Energy Internet

Smart electronics are beginning to make this scenario possible. When fuel cells are linked electronically with our existing power distribution systems, customers would be able to buy and sell energy freely. New software and advanced electronic systems could make it easier to connect all the pieces of our electricity network, enabling hydrogen-powered fuel cells to trade energy with the utility grid via plug-and-play connections.

In this vision, micropower in industrial countries does not displace utility power, which would still be needed during a long transition, but would embrace the grid and grow along with it. The grid would use new technologies for intelligent metering and switching, and ways could be invented to control the stream of electrons. Eventually, electricity grids might be transformed "from dictatorial monopolies into democratic marketplaces," says Vaitheeswaran.[20]

Jeremy Rifkin, in his book *The Hydrogen Economy*, also envisions more open energy markets and a fusion between hydrogen energy and the ongoing revolution in computers and telecommunications. If renewable energy is harnessed, "every human being on Earth could be 'empowered,' making hydrogen energy the first truly democratic energy regime in history."[21]

The biggest future market for hydrogen fuel cells and other micropower plants could be the developing world. Building up a massive electricity infrastructure in these countries is not affordable, but renewable energy is often abundant and could be used to make hydrogen for micropower plants.

Eventually, everyone might be connected to an energy internet or hydrogen energy web. The whole system could be managed by virtual utilities, and fuel cells could monitor energy prices over an internet connection or via digital signals embedded in the electricity itself. Individual cars might become "the roaming palmtops of the Energy Web."[22]

To realize this concept, scientists and engineers are working on technologies that would enable a gradual, global transition toward hydrogen energy systems. Think long-term, though. Depending on how old you are now, you might not see hydrogen power become practical in everyday life, but maybe your children or grandchildren will.

References

1. Lovins, Amory, in Vaitheeswaran, Vijay V., *Power to the People: How the Coming Energy Revolution Will Transform an Industry, Change Our Lives, and Maybe Even Save the Planet.* Farrar, Straus and Giroux: New York, 2003. ISBN 0-374-23675-5, page 240.

2. Vaitheeswaran, page 317.

3. Hoffman, Peter, *Tomorrow's Energy: Hydrogen, Fuel Cells, and the Prospects for a Cleaner Planet.* The MIT Press: Cambridge, MA, 2001. ISBN 0-262-08295-0, page 8.

4. U.S. Department of Energy, Office of Energy Efficiency and Renewable Energy, DOE *Hydrogen Posture Plan,* February 2004, www.eere.energy.gov, page iii.

5. Naki´cenovi´c, Nebojsa, in Rifkin, Jeremy, *The Hydrogen Economy: The Creation of the Worldwide Energy Web and the Redistribution of Power on Earth.* Penguin Group (USA) Inc.: New York, 2002. ISBN-1-58542-193-6, page 178.

6   Lovins, Amory, in Vaitheeswaran, page 240.

7. Vaitheeswaran, pages 318-319.

8. Hoffmann, page 9.

9. Vaitheeswaran, page 242.

10. Hoffmann, page 6.

11. Rocky Mountain Institute, www.rmi.org.

12. Ibid.

13. U.S. Energy Information Administration, *Annual Energy Outlook 2004 with Projections to 2025,* www.eia.doe.gov/oiaf/aeo.

14. Yeager, Kurt, Electric Power Research Institute, in Silberman, Steve, "The Energy Web," *Wired,* Issue 9.07, July 2001, www.wired.com.

15 Vaitheeswaran, page 20.

16. General Motors Corp. www.gm.com, 2000.

17. BP plc, www.bp.com.

18. Lovins, Amory, in Rifkin, page 208.

19. Vaitheeswaran, page 16.

20. Ibid., page 43.

21. Rifkin, page 9.

22. Ibid., pages 9, 199; Vaitheeswaran, page 43; Silberman, Steve, "The Energy Web," *Wired,* Issue 9.07, July 2001, www.wired.com.

Drivers and **Barriers**

Just a few years ago, American automakers could hardly give away an all-electric car, despite generous tax credits. California had forced manufacturers to offer zero-emission vehicles, but customers roundly rejected them.

Yet today, people are eager to join months-long waiting lists to buy hybrid gasoline-electric vehicles, and automakers are tacking several thousand dollars onto their prices. In this case, government policy didn't work to mandate a market for an environmentally beneficial product—instead, technology advances and customer demand created it.

Several interrelated forces are driving the world's industrial nations toward a hydrogen economy that could use fuel cells for much of its electricity, from multi-megawatt plants down to home generators and car engines. Eventually our supplies of hydrogen could be provided by renewable, pollution-free sources. What are the most important forces driving us toward this vision, and what are the chances that hydrogen energy systems will become a reality?

Although our dependence on oil imported from the Middle East is often cited as a reason to develop alternative energy, environmental concerns are just as crucial, given the mounting evidence of global warming. As an added propellant, energy markets have become less constrained by regulation, especially in America and Europe, allowing greater opportunities for entrepreneurs. And technological innovation is unleashing a wave of improvements in the cost and convenience of energy services and products.

Chief among these improved technologies are fuel cells, which could launch hydrogen into the world's energy mix. Fuel cells and other micropower applications are enjoying a renaissance, due partly to the difficulty and cost of expanding our conventional electric power infrastructure. Micropower (also called distributed generation) could provide the higher quality, more reliable electricity required by our increasingly computerized economy.

The automotive industry is pioneering the development of practical, desirable hydrogen fuel cell-powered vehicles, joined by oil companies and equipment manufacturers. Governments worldwide are funding research, often via public-private partnerships aimed at developing products that will succeed in the marketplace. Demonstrations of hydrogen energy in all kinds of applications are taking place around the globe with widespread support.

But there are some clouds on the horizon. Environmental advocates find fault with some pathways toward a hydrogen future, most notably production of hydrogen from coal and nuclear power or even from natural gas. Fuel cells could still stumble badly on the road toward mass production. And a few years of low oil prices could distract efforts to increase our energy independence, which many analysts say is a questionable venture anyhow.

Some scientists conclude that the prospects for a hydrogen economy are uncertain, despite many advantages in its favor. In *Issues in Science and Technology*, Daniel Sperling and Joan Ogden express the need to cultivate a shift toward hydrogen:

> The transition to a hydrogen economy will be neither easy nor straightforward. Like all previous alternatives, it faces daunting challenges. But hydrogen is different. It accesses a broad array of energy resources, potentially provides broader and deeper societal benefits than any other option, has no natural political or economic enemies, and has a strong industrial proponent in the automotive industry.

In the end, though, the hydrogen situation is precari-
ous. Beyond a few car companies and a scattering of entrepreneurs, support for hydrogen is thin. Although many rail
against the hydrogen hype, the greater concern perhaps
should be the fragile support for hydrogen....It appears
to us that hydrogen is a highly promising option that we
should nurture as part of a broader science, technology,
and policy initiative. The question is how, not if.[1]

Other observers also cherish the hope that we can move toward a
cleaner, smarter, more sustainable energy future. In *Power to the People*, Vijay
Vaitheeswaran pins his optimism on micropower:

Stopping the use of fossil fuels completely and immediately would be foolish and needlessly expensive, but
a thoughtful, phased shift to hydrogen-fired micropower
would not. On the contrary, the innovative technologies
unleashed by market liberalization and environmental demands hold out the promise of an inexpensive, and maybe
even profitable, transition to a cleaner energy world. If we
grasp that opportunity, then there is every reason for hope
about our planet's future. Indeed, there is every reason to
think that today's nascent energy revolution will truly deliver power to the people.[2]

Many different factors, from the desires of wealthy consumers to the
plight of the world's poor, are attracting international attention to hydrogen
and stimulating business and government to invest in research and development. Here we'll take a look at the reasons behind this unprecedented attention, and we'll identify the jumble of proponents—some of them traditional
rivals—who are getting involved in a big way.

Our planet's health

Hydrogen's potential to mitigate air pollution has engaged scientists for decades, and their mission was galvanized by the environmental movement of the 1960s and 1970s. After more or less languishing in the 1980s, environmental activism was rejuvenated in the 1990s as reports of global warming emerged, blaming greenhouse gases such as carbon dioxide (CO_2) and methane (CH_4) for an apparent trend toward a hotter planet.

Other kinds of air pollution also persist in many areas, including acid rain, caused by oxides of nitrogen and sulfur (NO_x and SO_x), and ground-level ozone, formed from NO_x and volatile organic compounds (VOCs), as well as generous helpings of smoke and soot (particulates).

"Environmental pollution will likely represent the 'cold war' of the next century." This idea was suggested to the automotive industry at a fuel cell workshop in 1998 by Alan C. Lloyd, who was soon thereafter appointed Chairman of the California Air Resources Board.[3]

Making hydrogen from natural gas and other fossil fuels would involve some amount of air pollution, depending on the production scale. Many environmental activists would produce hydrogen only from renewable sources like wind power and solar energy. Europe and Japan have increased their renewable energy use, and wind turbine capacity has grown rapidly in Europe and America for the past several years—trends that bolster the potential for hydrogen to help control pollution.

The theory of global warming, also called climate change, is still extremely controversial—confusion flourishes, simple answers elude us, and our headline-driven news doesn't always help clarify the issue.

But given the most recent findings from objective scientists, we can't deny that at some point in the future, we will have to get our energy in ways that release fewer greenhouse gases. Given the scientific uncertainties, climate change could result in "unpleasant, unpredictable surprises."[4] Hydrogen could help us dodge these surprises, or at least diminish their impact.

Potential effects of climate change

The effects of global warming might benefit some frigid northern countries, which could enjoy more moderate weather. But the tropics would bear the brunt of climate change, and that's where most of the world's poorest populations live. Unfairly, these people produce relatively little air pollution themselves.

Also, in continents with a huge north-south spread, the effects overall might seem relatively benign, but would mask regional extremes. In North America, for instance, we could see more frequent floods and droughts, longer heat waves, and more powerful storms, according to the nonpartisan Pew Center on Global Climate Change.[5]

However, to clear up one common misperception, climate change is a very gradual process, scientifically speaking. When people in New York City suffer through a week of sweltering weather, it's easy to blame global warming. Then when it's bitterly cold the next winter, they might say, "Where's global warming when we need it?" Who among us hasn't heard comments like this?

But weather conditions over periods of weeks, months, or even years don't by themselves prove any trend of climate change. Rather, scientists are collecting data covering hundreds of years, and they're using computer models to estimate changes that have occurred over thousands of years before that. They've even found ways to detect "human fingerprints" in the atmosphere.[6]

Evidence of global warming

Some effects of global warming are already apparent. In Alaska, where the permafrost has begun to soften, roads are crumbling and houses are sagging into their foundations.[7] At the North and South Poles—well-recognized bellwethers of environmental change—glaciers have begun melting, causing a measurable rise in sea levels.[8] Research suggests that crop yields of rice, a global staple, might be reduced by warm nighttime temperatures.

Whether these effects are due to human activities, such as burning fossil fuels, has been disputed for some time. Questions also have been raised about the validity of temperature data and the boundaries of what might be considered normal variability for our planet.

National Research Council. The National Research Council, which is part of the U.S. National Academies, explored these issues and determined that the earth's surface temperatures over the past 100 years increased by roughly 0.4°C to 0.8°C (0.7°F to 1.4°F), a rate "substantially greater than average." Global warming is "undoubtedly real," said the report.[9]

Although this finding clarified certain scientific squabbles about where and how temperatures had been measured, the National Research Council stopped short of linking the temperature increase to greenhouse gas emissions from human activities.

Georgetown University. Similarly, John McNeill of Georgetown University examined the causes and effects of global warming. He noted that while greenhouse gases in the atmosphere make the earth trap more heat from the sun, dust and soot slightly obstruct some amount of solar radiation reaching the earth.

The net effect since around 1800 has been an increase in the amount of solar energy delivered to the earth's surface of about 2 Watts per square meter (10 square feet). "This probably accounts for the modest warming the earth experienced in the twentieth century."[10]

McNeill quantified the past century's warming trend and identified surges in 1910–1940 and from 1975 to the present. Although average temperatures fell slightly in-between these surges, McNeill said:

> ...Nine of the ten hottest years on record occurred between 1987 and 1997, and the 1990s promised to be the hottest decade since the fourteenth century. Changes of this magnitude and rapidity are well within the natural range of variation, although rare within the last 2 million years, probably nonexistent within the last 10,000 years, and definitely absent within the last 600 years.

However, said McNeill, "no one knows for certain if human actions are the cause."

United Nations. Then more evidence of global warming—and a strong indictment of humanity as one of its causes—came from the United Nations' Intergovernmental Panel on Climate Change (IPCC) in 2001. The IPCC's scientists included skeptics of the global warming theory, so its conclusions are well-respected as being more objective than those from some other sources.

The IPCC estimated that if nothing is done to counteract global warming, the planet's temperature could increase within the next century by a range of 1.4°C to 5.8°C (2.5°F to 10.4°F), compared with 1.0°C to 3.5°C (1.8°F to 6.3°F) estimated in its previous report. "The stakes associated with projected changes in climate are high."[11]

Moreover, the IPCC said unequivocally that human activities have "contributed substantially" to global warming over the past 50 years—a conclusion that had been soft-pedaled in previous reports and questioned by many other scientists.

National Academy of Sciences. In America, the second Bush Administration wanted more proof and directed the NAS to investigate. The NAS panel also included skeptics, yet it agreed essentially with what the IPCC had found:

> Greenhouse gases are accumulating in Earth's atmosphere as a result of human activities, causing surface air temperatures and subsurface ocean temperatures to rise. Temperatures are, in fact, rising. The changes observed over the last several decades are likely mostly due to human activities, but we cannot rule out that some significant part of these changes is also a reflection of natural variability. Human-induced warming and associated sea level rises are expected to continue throughout the 21st century.[12]

European Environment Agency. The effects of global warming in Europe were examined by this agency in a report covering 22 indicators in eight categories: atmosphere and climate; snow, ice, and glaciers; marine systems; terrestrial ecosystems and biodiversity; water; agriculture; the economy; and human health.[13]

Compared with global temperatures, Europe's warming trend is slightly stronger, with an increase of 2.0–6.3°C (3.6–11.3°F) estimated for the next century. Weather extremes, glacial retreat, and rising sea levels were also quantified and projected.

Methane emissions

Methane, the prime constituent of natural gas, is a potent greenhouse gas that has been indicted, along with CO_2, as a cause of global warming. About half of the world's methane emissions are believed to come from five countries, according to the U.S. Environmental Protection Agency: China, Russia, India, the United States, and Brazil.

Methane is released from leaky natural gas distribution systems, coal mines, rice paddies, livestock, and municipal landfills. Unlike carbon dioxide, however, methane is a valuable product, and methods of harvesting the released gas could pay off. New technologies being developed include hand-held infrared video cameras that can detect methane leaks from a distance and display the gas as a visible cloud.

An international effort is under way to capture methane emissions. The United States plans to invest up to $53 million to promote methane harvesting over five years, working with Australia, India, Italy, Japan, Mexico, the United Kingdom, and Ukraine. Canada and Russia could join the program.

Government actions

Even before the conclusion of the global warming studies, the United Nations had initiated the Kyoto Protocol, a treaty that mandates reduction of greenhouse gas emissions. It uses 1990 as a starting point and calls for specific percentage decreases below that level.

The world's richest countries were the prime targets for signing the treaty since they produce the most greenhouse gases, especially the United States at five times the global average for CO_2.[14] Poor countries were motivated to sign by incentives that would promote clean energy investments in their economies.

The European Union agreed to overall reductions of 8%, with individual targets for member countries (Table 2–1), and Japan accepted a 6% decrease. More than 125 countries worldwide had signed or ratified the Kyoto Protocol as of mid-2004,[15] and Russia ratified it near year-end. Russia's approval was critical to reaching the protocol's requirement that signing countries account for at least 55% of global CO_2 emissions.

TABLE 2–1. EUROPEAN UNION CO_2 EMISSIONS TARGETS. EUROPEAN COUNTRIES HAVE AGREED TO REDUCE THEIR CO_2 EMISSIONS BY VARYING AMOUNTS, USING 1990 AS A BASELINE. TABLE COURTESY OF THE U.S. ENERGY INFORMATION ADMINISTRATION.

	Commitments (% change in emissions for 2008-12 relative to base-year levels)
Austria	–13
Belgium	–7.5
Denmark	–21
Finland	0
France	0
Germany	–21
Greece	+25
Ireland	+13
Italy	–6.5
Luxembourg	–28
Netherlands	–6
Portugal	+27
Spain	+15
Sweden	+4
United Kingdom	–12.5
Source: Council Decision 2002/358/EC	

Despite early progress, Europe could fall short of its goals unless new policies and measures such as emissions trading are implemented.[16] The European Climate Exchange, formed in 2004, will allow European companies to trade greenhouse-gas emissions credits. If a company exceeds its emissions limit, it can buy credits from others that have produced lower emissions than allowed. Futures trading and cash markets are planned.[17]

In America, President George W. Bush pulled out of the Kyoto agreement in 2001 and proposed an alternative, voluntary path to slow the growth of greenhouse gas emissions. This surprised some electric utilities and other industrial plant operators, who had been anticipating that CO_2 regulations would clarify future requirements so that they could invest confidently in new facilities. Since then, many large businesses have taken the initiative to reduce CO_2 emissions in the absence of federal rules, due in part to shareholder pressure.

In addition to political complications, estimates vary wildly on the future costs of combating global warming. What's more, we won't know for decades whether our efforts to control climate change are too strong or too weak. The computer models just aren't that exact, and the amount of change being projected is too tiny. How can we take such unprecedented actions when the problem is so ambiguous?

"That, in a nutshell, captures the dilemma of climate change," says Vaitheeswaran. "It is asking a great deal of politicians to think, let alone act, on behalf of voters who have not yet been born."[18] He urges that when the Kyoto treaty is revised, Europe and other governments should agree to start reducing emissions modestly rather than drastically. Also, America should acknowledge that the world is becoming carbon-constrained, and governments must step in directly to promote and invest in science and technology.

On a positive note, a similar global challenge was overcome, not that long ago, when scientists identified emissions of chlorofluorocarbons (CFCs) as a culprit in puncturing the earth's protective ozone layer. Among the parallels to today's global warming problem, a key to the solution then was participation by industry, even if selfishly motivated. That's why it's encouraging today to see automakers worldwide, and even some oil companies, involved in developing hydrogen technologies, regardless of their agenda.

Whether hydrogen will figure in the outcome of the global warming dilemma is still a wild card, but hydrogen's pollution-free nature can't be denied. As Peter Hoffmann puts it, "Switching to hydrogen energy—even perhaps to hydrogen [produced] from fossil fuels as a stopgap measure—may help save our children's health and perhaps their lives."[19]

Air pollution in the developing world

Carbon dioxide emissions in the developing world (including China and India) are expected to surpass those in industrial countries before 2025 (Fig. 2–1), said the U.S. Energy Information Administration (EIA) in its *International Energy Outlook* 2004. Even with this increase, however, developing countries will still account for less than half of global CO_2 emissions.[20]

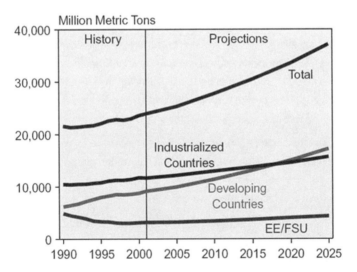

FIG. 2–1. GLOBAL CO_2 EMISSIONS BY REGION. BY 2020 DEVELOPING COUNTRIES WILL PRODUCE MORE CO_2 THAN THEIR INDUSTRIAL COUNTERPARTS EVEN THOUGH THEY USE LESS ENERGY. ARTWORK COURTESY OF EIA.

Besides containing no carbon, and thus forming no CO_2 or carbon monoxide, hydrogen does not generate other air pollutants associated with combustion of hydrocarbon fuels, such as SO_x, VOCs, and particulates, although hydrogen fuel can form NO_x when it's burned with air instead of oxygen. Except for large cities and other persistent pockets, these air emissions in Europe and America are being mitigated by regulations, but the developing world is another story.

Even though many poor countries possess huge solar or wind resources, most of their population has no access to electricity, and primitive fuels such as wood, charcoal, and cattle dung cause air pollution both indoors and out. "Hardest hit are women, whose responsibility it often is to provide fuel for household use," points out Steve Silberman in *Wired*. He cites statistics that document premature deaths in India and high respiratory illness rates among females in China, due largely to indoor air pollution.[21]

In addition to poor indoor air quality, the world's rapidly growing economies in southern Asia have created a two-mile-thick cloud of pollution called the Asian brown haze by Klaus Töpfer.[22] Beijing and other Chinese cities are working to control the spread of environmental pollution.

Hydrogen-fueled micropower could improve the lives of people dependent on primitive fuels. Many programs are under way to install solar and wind energy equipment in the developing world. Eventually, hydrogen fuel cells and other micropower stations could deliver energy services to the world's poor using their abundant renewable energy as a source of pollution-free hydrogen fuel.

Industry observer Jeremy Rifkin, in his book *The Hydrogen Economy*, foresees a world where the price of fuel cells could fall to the point where they'll be practical for installation in poor villages and cities. The use of local, renewable power to make hydrogen for fuel cells, which would then generate electricity on demand, would not only mitigate air pollution but also avoid the need to build a costly power transmission infrastructure, virtually from scratch in many areas of the globe.

Making the shift to a hydrogen energy regime—using renewable resources and technologies to produce the hydrogen— and creating distributed-generation energy webs that can connect communities all over the world is the only way to lift billions of people out of poverty. Narrowing the gap between the haves and have-nots means first narrowing the gap between the connected and the unconnected.[23]

Even the World Bank has embraced smart growth, which encourages environmental protection along with economic development. The bank estimates that China contains 16 of the world's 20 most polluted cities. Substantial pollution can be abated at low cost in poor countries, and it makes sense for them to tackle some environmental problems before people become wealthier.[24]

Investment in cleaner, more efficient energy technologies in developing nations will be promoted by the U.S. Overseas Private Investment Corp., under an agreement with the U.S. Department of Energy (DOE) in 2004. The two agencies will support U.S. investment in environmentally sound economic growth in emerging markets.

Efficiency as a route to pollution control

By any measure, hydrogen fuel cells are more efficient than traditional power generation plants based on combustion of fossil fuels and much more efficient than the lowly car engine. Efficiency matters—not just from the perspective of saving money, but also because consuming less energy reduces the emissions associated with burning hydrocarbon fuels.

Hydrogen fuel cells can be applied in our transportation sectors for automotive power and in stationary plants for power generation on just about any scale, from homes to utilities. In the United States in 2002–03, transportation accounted for about 33% of our CO_2 emissions (Fig. 2–2)—larger than any of the other three sectors (residential, commercial, and industrial), said EIA in its *Annual Energy Outlook* 2004.[25]

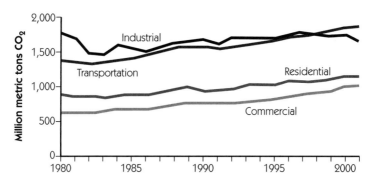

Fig. 2–2. U.S. CO$_2$ emissions. The transportation sector accounted for 33% of the U.S. total. Artwork courtesy of EIA.

In the U.S. transportation sector, CO$_2$ emissions are expected to grow until 2025 at an average annual rate of 1.9%, with the fuel economy of passenger vehicles increasing only slightly above 2002 levels.

In the electric power industry, the use of fossil fuels accounted for about 39% of total energy-related CO$_2$ emissions in 2002–03, and its share is projected to increase to 41% in 2025, said EIA (Fig. 2–3). Hydrogen fuel cell vehicles and micropower could help control emissions from the transportation sector and from electric power operations.

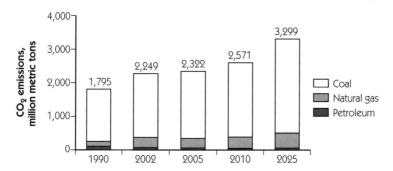

Fig. 2–3. U.S. electric power sector CO$_2$ emissions. This sector's fossil fuel consumption accounts for 39% of total energy-related CO$_2$ emissions, and its share is expected to rise to 41% by 2025. Artwork courtesy of EIA.

Fuel cell vehicle efficiency. Most vehicles today rely on the internal-combustion engine, which is a poor fit to the demands of a vehicle. Essentially, the engine works by harnessing continuous, tiny explosions—that's the process by which it converts the fuel's chemical energy into useful mechanical energy. Despite a century of advances in technology, most internal-combustion engines capture only 15% to 30% of the energy in gasoline. The rest is lost as waste heat, due to friction between moving parts, and vibrational noise.

In contrast, an electric motor is much better suited for vehicle propulsion because it delivers maximum torque just when the car needs it most, at low speeds. When the driver is braking or coasting down a hill, the electric motor, acting as a generator, captures the vehicle's momentum as electricity and uses it to charge the battery.

The problems of electric motor-driven cars—short traveling range, poor performance, and tedious battery-recharging—have been mitigated by today's generation of hybrid electric-gasoline vehicles. But tomorrow's breed of hydrogen fuel cell-powered vehicles could be even more efficient, as well as pollution-free.

Running on hydrogen and oxygen, the fuel cell generates electricity for the motor, and the vehicle emits nothing but water vapor. Hydrogen fuel cells in vehicles range in efficiency up to 60%. Hydrogen tanks already being demonstrated can be refueled in about five minutes, and some fuel cell vehicles can approach the traveling range of their gasoline-fueled counterparts (see chapter 4).

Fuel cell power plants. Central power generation plants are also relatively inefficient. Overall, the U.S. power system operates at a fuel-to-electricity efficiency of only about 33% to 35%. Smaller power generating units, such as modern natural gas-fired turbines and large engines, perform better from an efficiency and emissions standpoint. These can achieve efficiencies in the 40–50% range with relatively low emissions.

When their waste heat is recovered and used (for example, in combined cycles using two turbines), these cogenerating systems can reach 70% to 80% efficiencies. In comparison, fuel cells in stationary power generation applications operate at efficiencies up to 65%, and using cogenerated heat, a fuel cell system can approach efficiencies of 85%.

Profiting from air pollution control?

Most politicians and economists—and many environmentalists—believe that solving the climate problem will be financially painful. They assume that CO_2 and other greenhouse gases are the inevitable result of economic activity, and so any attempt to reduce these emissions would hurt the economy.

However, according to the Rocky Mountain Institute, efficiency is the key to cost-effective air pollution control. The best climate-protection measures actually help the economy and offer profitable business opportunities. Simply implementing cost-effective energy efficiency—that is, measures that pay a better-than-market rate of return—could eliminate over half the threat of global climate change.[26]

This strategy worked for oil giant BP, which ended up making money when it decided to reduce CO_2 and methane emissions from its operations by 10% below 1990 levels. Its original goal was to do this by 2010—but the target was achieved nine years ahead of schedule in 2001.

"We found that efficiency and emission reduction was good business," said Lord Browne, BP's group chief executive. "In fact within the first three years we added $650 million of value, for an investment of around $20 million." One of the reasons was, he said, "The aspiration resonated with the expectations of our staff. Doing something positive for the global environment generated enormous enthusiasm and creativity."[27]

Another optimistic viewpoint regarding environmental pollution comes from Jesse Ausubel of Rockefeller University. As the world's developing nations gradually grow wealthier, he believes, the process will be slow enough to allow them to implement new, more efficient technologies along the way, which will diminish the impact of their increased economic activity on the environment. For example, he says, "When China has today's American mobility, it will not have today's American cars."[29]

A good example of this leapfrog effect is telephone communications—developing countries aren't wasting money building a land-line infrastructure, they're going straight to cellphones. There are limits to how much technology can accomplish, however, and innovation does not always reduce pollution. Governments would still have to encourage the kind of technologies that abate pollution rather than exacerbate it.

Examples of technological innovation on the hydrogen front abound—not only in fuel cells and other micropower appliances, but also in methods of producing, storing, transporting, and distributing hydrogen (see chapters 4–7).

The Global Oil Picture

One of the most attractive features of any hydrogen fuel cell car is that it does not rely on "the iron nexus of gasoline and the internal-combustion engine," as described by Vijay Vaitheeswaran. Although fuel cell vehicles are a long way from mass production, they could eventually spell the doom of the oil age.[30]

The crux of the issue is not just the amount of oil left for us to exploit, but also its location, primarily in the Persian Gulf. The hydrogen needed for fuel cells can be produced anywhere in the world, in many different ways, and from just about any energy source. The world's industrial economies would no longer depend on the Middle East for most of our energy. We could "never be held hostage by a future Osama bin Laden."[31]

Oil consumption trends

Hydrogen enthusiasts, who have thrived for decades in the nether-regions of the world's research facilities, vaunted to the forefront in the 1970s when an embargo by the Organization of Petroleum Exporting Countries (OPEC) announced the end of cheap oil. Since then, we have learned that depletion of the world's supply will not happen as soon as we thought, yet it's still an issue of some urgency, depending on whose figures you believe.

Prodded by the specter of an oil-less world, industrial nations have dramatically reduced their energy consumption per unit of economic output (gross domestic product, or GDP) since the 1970s. In other words, businesses and factories worldwide now use much less energy to produce the same amount of goods and services. As an example, the United States slashed its energy consumption per dollar of output by almost 50% from 1970 to 2002, according to EIA (Fig. 2–4).

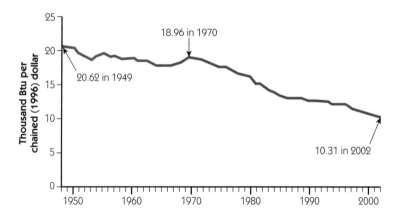

FIG. 2–4. U.S. ENERGY USE. SINCE 1970, THE UNITED STATES HAS CUT ENERGY CONSUMPTION PER DOLLAR OF ECONOMIC OUTPUT ALMOST IN HALF. ARTWORK COURTESY OF EIA.

Nevertheless, global oil consumption continues to climb, driven not just by energy-wasting habits but also by the sheer volume of people in rapidly developing countries and their increasing demand for oil as their economies modernize.

This highlights the importance of conservation and efficiency in extending the life expectancy of our global oil supplies. According to mathematician Evar Nering of Arizona State University, curbing demand is more effective than boosting supply. Statistically, cutting annual demand growth in half (from 5% to 2.5%) would make our existing oil supply last twice as long. In contrast, doubling our existing reserves of oil would add only 14 years to the life of our current supply, if growth in demand is unchecked.[32]

Today global oil demand is growing rapidly, not only in the United States—by far the world's biggest oil consumer—but also in developing nations in Asia, South America, and Africa (Fig. 2–5). Through 2025, China's oil consumption is projected to grow at 4.8%/yr and India's by 4.7%/yr. In terms of overall energy consumption, developing Asia accounts for 40% of the increase through 2025 (Fig. 2–6), said EIA's *International Energy Outlook* 2004.

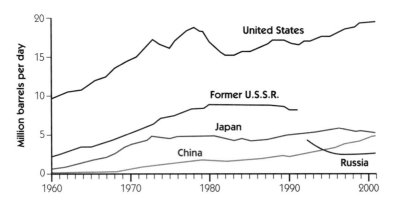

FIG. 2–5. GLOBAL OIL CONSUMERS. OIL USE CONTINUES TO GROW WORLDWIDE, LED BY CHINA—WHICH SURPASSED JAPAN AS THE WORLD'S SECOND-LARGEST CONSUMER IN 2003. ARTWORK COURTESY OF EIA.

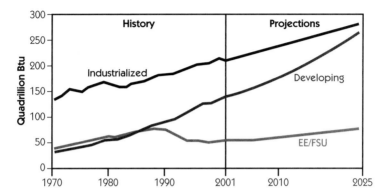

Fɪɢ. 2–6. Eɴᴇʀɢʏ ᴜsᴇ ɪɴ ᴅᴇᴠᴇʟᴏᴘɪɴɢ ɴᴀᴛɪᴏɴs. Sᴛʀᴏɴɢ ᴇᴄᴏɴᴏᴍɪᴄ ɢʀᴏᴡᴛʜ ɪɴ Cʜɪɴᴀ, Asɪᴀ, ᴀɴᴅ ᴏᴛʜᴇʀ ᴅᴇᴠᴇʟᴏᴘɪɴɢ ʀᴇɢɪᴏɴs ᴡɪʟʟ ɢʀᴇᴀᴛʟʏ ɪɴᴄʀᴇᴀsᴇ ᴛʜᴇɪʀ ᴅᴇᴍᴀɴᴅ ꜰᴏʀ ᴇɴᴇʀɢʏ. Aʀᴛᴡᴏʀᴋ ᴄᴏᴜʀᴛᴇsʏ ᴏꜰ EIA.

Developing countries run the risk of getting just as hooked on oil as their rich counterparts. And when oil prices spike, their economies suffer and their debt burdens mount. Hunting for new oil reserves is obviously worthwhile, but conservation and efficiency measures are needed in industrial and developing nations alike.

How soon will we run out of oil?

As in the global warming debate, scientific experts argue heatedly about when the world's oil production will peak and ultimately be depleted. Some analysts think we've already produced as much oil as we're ever going to, while others project the peak in the 2030s, 2040s, or beyond.

A series of articles by *Oil & Gas Journal* in 2003 and 2004 examined this issue in depth. Oil industry experts fall into the imminent peak group, which contends that the world is now at or near its maximum oil production levels, versus those who believe that steady growth in reserves will delay the point when we reach that peak. According to three forecasts of world oil production, the peak will occur no later than 2016 (Fig. 2–7).

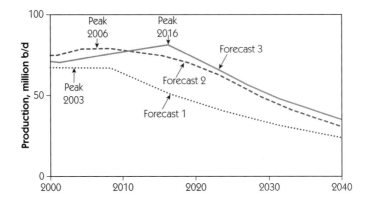

Fɪɢ. 2–7. Fᴜᴛᴜʀᴇ ᴡᴏʀʟᴅ ᴏɪʟ ᴘʀᴏᴅᴜᴄᴛɪᴏɴ. Aᴄᴄᴏʀᴅɪɴɢ ᴛᴏ ᴛʜʀᴇᴇ ꜰᴏʀᴇᴄᴀꜱᴛꜱ, ᴡᴏʀʟᴅ ᴏɪʟ ᴘʀᴏᴅᴜᴄᴛɪᴏɴ ᴡɪʟʟ ᴘᴇᴀᴋ ʙʏ 2016. Aʀᴛᴡᴏʀᴋ ᴄᴏᴜʀᴛᴇꜱʏ ᴏꜰ *Oil & Gas Journal.*

This view foresees annual production declines of about 2.5% over the 20 years following the peak. This post-peak decline rate is important, since it determines the time available for transition to a new energy regime such as hydrogen. A more gradual slide in production over several decades would allow a more manageable transition.

Many analysts count on discoveries and development of new oil fields to increase reserves substantially. However, a large increase in the amount of recoverable oil we find would delay the peak in world oil production by only a few years, according to the Institute on Energy & Man.[33] Nevertheless, this additional oil could significantly slow the rate of decline after the peak is reached—which is more important to achieving a gradual transition toward other fuels (Fig. 2–8).

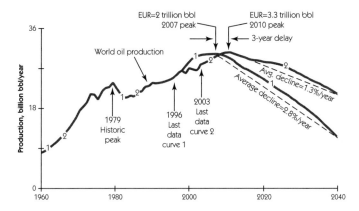

FIG. 2–8. PEAK OIL DELAY. AN INCREASE OF 3.3 TRILLION BARRELS IN THE
WORLD'S ESTIMATED ULTIMATE RECOVERY (EUR) WOULD DELAY THE OIL
PRODUCTION PEAK BY ONLY A FEW YEARS, BUT COULD SIGNIFICANTLY REDUCE THE
RATE OF DECLINE AFTERWARDS. ARTWORK COURTESY OF *Oil & Gas Journal.*

The U.S. EIA and many oil companies are more optimistic about the
world's oil output. Using U.S. Geological Survey data, EIA projects healthy
growth in production of conventional oil through 2020 and a peak around
2037–2047 (Fig. 2–9).

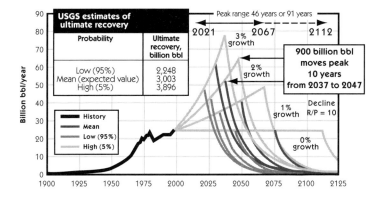

FIG. 2–9. CONVENTIONAL OIL PRODUCTION FORECASTS. DEPENDING ON
GROWTH IN WORLD OIL RESERVES, GLOBAL PRODUCTION WILL NOT PEAK FOR
MANY DECADES. ARTWORK COURTESY OF *Oil & Gas Journal.*

Dependence on imports from the Middle East

Even if we do have decades left before we must end our addiction to oil, the other problem is its source, and on this issue there's very little debate. The Middle East commands more than 40% of the world's oil, most of it around the Persian Gulf (Fig. 2–10).

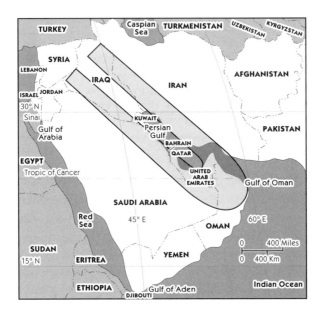

FIG. 2–10. PERSIAN GULF OIL RESERVES. ABOUT 90% OF MIDDLE EASTERN OIL RESERVES SURROUND THE PERSIAN GULF, WITH SAUDI ARABIA CONTAINING THE LARGEST AMOUNT (ABOUT 16% OF THE WORLD'S TOTAL). ARTWORK COURTESY OF Oil & Gas Journal.

As other big oil producers like Russia and the United States peter out in the future, the Middle East's share of global production becomes proportionately larger. The extent of this was documented by the National Iranian Oil Co. (Table 2–2).[34]

TABLE 2–2. MIDDLE EAST SHARE OF PRODUCTION. THE PERCENTAGE OF WORLD OIL PRODUCTION FROM MIDDLE EASTERN COUNTRIES COULD GROW TO NEARLY 48% BY 2020. ARTWORK COURTESY OF *Oil & Gas Journal*.

	2000	2005	2010	2020
		Million b/d		
World*	63.26	59.48	59.61	46.04
Middle East	20.83	18.84	23.17	22.03
Ratio ME/world, %	32.9	31.7	38.9	47.8

*The 64 major oil-producing nations.
Source: Colin Campbell, Association for the Study of Peak Oil database for conventional oil endowment, February 2002

Several trends have raised spirits that western countries might depend less on OPEC oil in the future. Supplies from a now-friendlier Russia have swelled, hindered only by its capacity to export the oil. Central Asian countries like Kazakhstan have found enormous new oil fields. And offshore west Africa and Brazil have seen some huge oil strikes.

But the numbers make clear that even these new supplies will pale in comparison to Middle Eastern oil resources. National Iranian Oil projects an irreversible decline in production beginning in 2006–07, due mainly to falling output from countries outside the Persian Gulf (Fig. 2–11).[35]

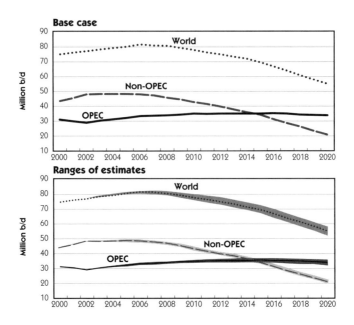

F$_{IG}$. 2–11. O$_{IL}$ production outlook. S$_{UPPLIES}$ produced from sources outside of OPEC have plateaued and could begin declining in 2006–07. A$_{RTWORK}$ courtesy of *Oil & Gas Journal*.

Can industrial nations produce and conserve their way out of the box and import less oil, period? The short answer is, not likely. Japan has virtually no oil resources, Europe's large North Sea fields are mostly mature, and U.S. crude oil production has recently hovered at 50-year lows.

In fact, America's dependence on imports will grow from 53% of its oil consumption in 2002 to a projected 70% in 2025, according to EIA (Fig. 2–12). This forecast includes domestic production from Alaska's North Slope but not from the Alaskan National Wildlife Refuge (ANWR).

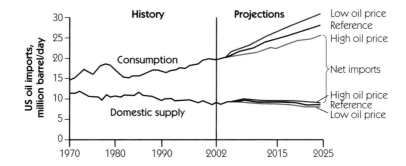

FIG. 2–12. U.S. OIL IMPORTS. AMERICA IS PROJECTED TO IMPORT 70% OF ITS OIL REQUIREMENTS BY 2025. ARTWORK COURTESY OF EIA.

Greater reliance on natural gas instead of oil has also been cited as a mitigating factor in the world's energy situation. Indeed, global gas reserves are growing, and trade in the form of liquefied natural gas (LNG) is thriving.

But gas resources in industrial nations are not in much better shape than oil supplies (Fig. 2–13), and the world's output of conventional gas might peak as early as 2020.[36] Japan has almost no gas, Europe's large fields are mature, and the United States will rely more on gas imports in the future (Fig. 2–14), with increases in consumption being led by the power generation sector.

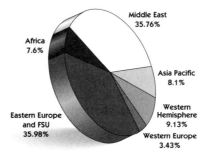

FIG. 2–13. NATURAL GAS RESOURCES. WESTERN COUNTRIES CONTAIN VERY LITTLE OF THE WORLD'S REMAINING GAS, AND GLOBAL OUTPUT OF CONVENTIONAL GAS MIGHT PEAK AS EARLY AS 2020. ARTWORK COURTESY OF Oil & Gas Journal.

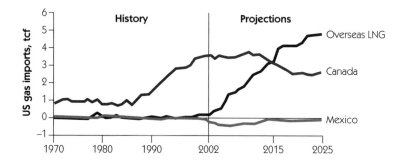

Fig. 2–14. U.S. natural gas imports. America will have to import more natural gas from overseas to feed new power plants and keep up with growing demand in other sectors. Artwork courtesy of EIA.

Energy security and geopolitics

After years of conflict between Israel and its Arab neighbors, the U.S.-led war in Iraq has introduced democracy there, but has not resolved the globe's geopolitical tensions. Terrorist attacks and sabotage continue and have threatened Saudi Arabia's oil industry, the heart of the world's supply.

Saudi Arabia is the only country, even in the Middle East, with substantial extra oil production capacity, so it serves as the swing producer in charge of rapidly ramping up deliveries and moderating prices. A successful strike at the Saudis' largest oil-processing complex might take millions of barrels of our daily oil supply out of the market.

The cost of protecting these foreign oil sources is huge. The terror premium on the price of a barrel of crude is estimated at one-fifth to one-fourth of the market price. The U.S. cost of the war in Iraq and the country's reconstruction exceeded $150 billion in 2004 and was estimated to reach $200 billion by 2006, on top of years of costly operation of military bases throughout the Mideast region.

Yet our efforts to break free from oil's grip on our industrial economies have largely failed. In part, this might be due to relatively low oil prices. Even $50 per barrel is nowhere near previous records in real terms (adjusted for inflation), and OPEC doesn't want higher prices.

"Oil prices that are too high...encourage the development of alternative energy sources," said Adel al-Jubeir, foreign-policy adviser to Saudi Crown Prince Abdullah.[37] So hydrogen and other alternative fuels are competing with a moving target, which is being purposefully manipulated.

Also, conveniently for OPEC, America and other rich countries tend to lose their political will when oil prices fall. As Leonardo Maugeri of Italy's energy giant Eni SpA observed, "Western countries have been historically unable to sustain a long-term foreign policy designed around energy objectives, which vanish once prices drop and [which] often conflict with broader diplomatic goals."[38]

Some political analysts question whether energy independence *per se* is a worthwhile goal for public policy. According to Holman W. Jenkins Jr., it's a "fantasy," at least in America. Even if the United States used less Middle Eastern oil, it would not be able to abandon its regional military and security entanglements and would still pay international prices for whatever oil we do buy.

Nor would U.S. actions financially harm Middle Eastern purveyors of oil, who would find ready markets elsewhere. Instead, he says, the real concern is price and its volatility—an issue that's resolving itself as industrial nations gradually wean themselves off oil.[39]

But that's no excuse to do nothing. As Jeremy Rifkin points out, "Rarely do societies respond to an 'anticipated' change in their circumstances. But when that potential change can radically affect the totality of our way of life and the very geopolitics of the world we live in, the collective nonchalance becomes a prescription for disaster."[40]

An opportunity for hydrogen

At the end of the day, it doesn't matter how much oil we have left, who has it, or how long it will last, many people agree. "The Stone Age did not end for lack of stone, and the Oil Age will end long before the world runs out of oil," said Sheik Zaki Yamani, whose words have been repeated by and attributed to many other speakers.

Rather, we should take advantage of whatever time is left to smooth the path toward hydrogen energy systems and alternative fuels. "The actual date of empty wells is largely irrelevant," says the Rocky Mountain Institute. "The many benefits of hydrogen will make petroleum fuels obsolete at low prices before their scarcity sends drilling costs skyward. In the coming years, we will begin to see our energy economy, now rooted in fossil fuels, replaced by a hydrogen economy."[41]

Big Business Buys In—Automaker/ Oil Producer Alliances

The world's richest countries have many reasons to shift toward hydrogen energy systems—to slow down global warming and control air pollution, to improve efficiency and conserve energy, to reduce oil consumption and mitigate dependence on imports, and to take advantage of technological innovations that meet consumer needs.

Now two of the world's biggest business sectors—automotive manufacturing and energy—have joined the hydrogen revolution and are supporting demonstration projects around the globe (see chapter 4). The world's governments are working with industry to fund research and development of fuel cells and other hydrogen systems. Electric utilities and equipment manufacturers are also getting on board. The participation of these players could make or break the success of hydrogen in tomorrow's energy systems.

The march of fuel cells toward the marketplace, and the chances of hydrogen energy becoming an everyday reality, have been propelled by what Vijay Vaitheeswaran calls a "happy collision" of forces—environmental concerns and popular activism, the loosening of energy market constraints, and technological innovations of fuel cells themselves. "Despite decades of frustration and failure, fuel-cell fans are once again jumping for joy."[42]

Automotive manufacturers have at times been at odds with the oil industry, partly due to each side's resistance to regulation in favor of the other. Automakers don't want to be forced to increase their vehicle's fuel economy or reduce tailpipe emissions to accommodate conventional gasoline. Oil companies don't want to be forced to provide special gasoline blends tailored to local and regional requirements or automakers' needs to meet standards.

So why have these two rivals joined hands in developing fuel cell-powered vehicles, hydrogen fueling stations, and other hydrogen systems? Automakers see hydrogen fuel cells as the next step in the technological evolution of vehicles, and each major manufacturer wants to be first to market. "Hydrogen's future appears to be tightly linked to automaker commitments to move fuel cells from the lab to the marketplace," said Sperling and Ogden in *Issues in Science and Technology*.[43]

The oil industry's vantage point is different, but still supportive of hydrogen to some degree. Today's oil companies are actually energy conglomerates that produce not just crude oil but petroleum products, natural gas, and other fuels and chemicals. The industry successfully battled the introduction of electric battery-powered cars because oil companies saw no future in this product for themselves.

But hydrogen does offer routes for the oil industry to profit. Energy companies are eager to supply any fuel that their customers might want. "Oil companies will not allow the hydrogen economy to develop without them," said Sperling and Ogden.

The oil/energy industry pioneers in promoting hydrogen are BP and Shell, both based in Europe, and America's ChevronTexaco. They have joined with several of the world's major automakers in demonstrating hydrogen fuel cell vehicles and fueling stations (see chapter 4). Other U.S.-based multi-national oil companies are generally less receptive to hydrogen and fuel cells, although some are taking part in demonstration projects.

The oil industry would not enjoy a rapid payback on hydrogen invest-ments, and the transition would have to be slow enough for companies to protect their existing assets. Government funding would be essential during the early stages of building a hydrogen infrastructure. Sperling and Ogden characterize the oil industry as "watchful, strategically positioning them-selves to play a large role if and when hydrogen takes off."

Changes in the Electric Power Industry

Unlike the transportation sector, most of the world's large power plants run on coal or nuclear energy, which are usually reliable, inexpensive sources of baseload electricity, our primary supply around the clock. However, prob-lems with the equipment and systems we use to deliver electricity to cus-tomers are driving demand for micropower—small, distributed generators that operate on or near the customer's property. These systems are more efficient than most central utility plants, and they avoid the loss of energy during long-distance power transmission, which is only about 5%–8% in the United States but up to 30% in developing countries.[44]

Hydrogen fuel cells are one of the most promising types of micropower, by virtue of their efficiency and cleanliness. Internal-combustion engines can be modified to accept hydrogen or hydrogen-methane fuels, and eventually microturbines and large turbines also could operate on hydrogen. Wind and solar systems also qualify as micropower, but they generate electricity inter-mittently, which makes them less practical. The cost of some micropower technologies has fallen substantially over the past decade, making them more economical for homes and businesses to buy and operate.

Restructuring and competition

Worldwide, governments of industrial nations have been gradually loosening regulations on electric utilities, encouraging or even forcing them to move from monopoly status into the marketplace as competitors. In the 1980s, non-utility companies that produce power independently were allowed to sell electricity to large customers and even to the utilities them-selves. The utilities also were required to let third parties use their transmis-sion lines at competitive rates (a practice called *wheeling*).

This essentially deregulated the power generation segment of the in-dustry and allowed the widespread creation of wholesale electricity markets, as well as retail markets in some areas. Many utilities spun off their own independent, unregulated businesses and sold their unprofitable power plants. Some utilities are also selling their transmission assets, and for-profit companies are springing up that are willing to build and maintain new transmission lines.

Restructuring of the electric industry has already spurred construction of distributed power plants in larger sizes (20 to 50 megawatts), primar-ily using natural gas-fueled turbine generators to supply both heat and power to nearby customers. These large cogenerating plants, along with emergency backup generators, account for 95% of customer-owned power capacity in the United States.[45]

Gradually, this trend should continue down the size scale as smaller plants become more economical to build and operate and as regulatory and institutional barriers to their construction are overcome (see *Barriers and Challenges* in this chapter).

Although the road toward competitive electricity markets has been far from smooth, the trend is clear, according to Vaitheeswaran: "The era of monopolization, centralization, and overregulation has started to give way to market forces in electricity. That, in a nutshell, explains why micropower has once again been given the chance to blossom."[46]

Demand for higher quality power

Many businesses consider power quality and reliability to be just as important as energy costs, due to the rapidly growing use of computers in 24/7 operations. However, several factors have increased the unreliability of electric grid networks, such as rising demand during peak hours—daily and seasonal times of maximum consumption, when electricity prices spike—and a lack of financial incentives to maintain or improve aging equipment. Transmission investment in the United States was $2.5 billion lower in 2000 than in 1975, while electricity sales nearly doubled in the same period.[47]

But the cost of building new transmission lines and power distribution systems has escalated, along with public opposition to such projects due to concerns such as electromagnetic fields. This has frustrated utilities that desperately need more power delivery capacity and has slowed the growth of competitors seeking to enter the transmission business.

Reliability and quality issues. Most customers perceive reliability as the absence of power outages, or unanticipated blackouts. As the world was reminded in mid-2003, our supply of electricity can vanish in an instant. More than 50 million people throughout eastern Canada and the midwestern and eastern United States lost power in late summer when a series of glitches triggered widespread plant shutdowns. Subsequent outages also occurred later that year in Europe on a smaller, but significant scale.

The Canada–U.S. blackout cost a total of around $6 billion, according to the U.S. DOE.[48] Other estimates range from $4.5 billion to $10.3 billion. New York City alone lost an estimated $1.1 billion, or roughly $36 million per hour.[49]

Other traditional reliability issues are rolling blackouts, which are intentionally imposed when loads exceed generating capability, and brownouts, a persistent lowering of voltage caused by overloading.

For many businesses, however, reliability also means high power quality—the absence of voltage surges, sags, or swells; impulse events (spikes or transients); voltage oscillations; harmonic distortions; and other split-second deviations. These minor quality disturbances can wreak havoc in many types of sensitive equipment, such as semiconductor microchip fabrication tools and electronic variable-frequency motor drives, which in turn can disrupt critical processes and control systems.

Annually, power outages cost the U.S. economy anywhere from $104 billion to $164 billion in lost productivity and other expenses. Power quality disturbances cost an additional $15 billion to $24 billion per year.[50] Globally, the market for premium power to protect customer equipment from these events is worth some $10 billion per year.[51]

Racking up the nines. Utilities do a good job of providing communities and businesses with power around the clock, which is available 99.99% of the time. On an annual basis, however, the 0.01% of the time that utility electricity might be unavailable translates to 53 minutes per year, an unacceptable level for many global operations. The cost of downtime is simply too high for today's 24/7 businesses (Table 2–3).

TABLE 2–3. OUTAGE COSTS FOR SELECTED COMMERCIAL CUSTOMERS

Industry	Average Downtime Cost
Cellular communications	$41,000/hr
Telephone ticket sales	$72,000/hr
Airline reservations	$90,000/hr
Credit card operations	$2,580,000/hr
Brokerage operations	$6,480,000/hr

Source: U.S. DOE

As homes and industries worldwide become ever more dependent on computers and other sensitive devices, our digital economy will increasingly require digital quality electricity—that is, always perfect and always on.

Future micropower systems like hydrogen fuel cells could deliver electricity 99.9999% of the time, with only 32 seconds of downtime per year not counting scheduled maintenance. This is called reaching the six nines of reliability (or six-sigma). Uninterruptible power supply (UPS) systems with multiple layers of redundancy can also deliver six nines, but at very high cost.

Even if the utility grid could achieve six nines of reliability, its power quality might still be poor in terms of voltage sags or other disruptions. In areas where upgrading or expanding the distribution network is too expensive, hydrogen fuel cell micropower can augment the grid supply, especially during peak hours. Early adopters of micropower systems such as fuel cells have included banks, hospitals, and office buildings (see chapter 4).

Meeting peak demand

Over the next two decades, America's electric utilities and other power providers will need to add baseload generation capacity, especially during 2010–2025, to meet rising demand in the commercial sector—the fastest-growing, at 2.2%/yr—and in the industrial sector as well, where consumption is expected to increase 1.6%/yr, according to EIA.

But residential demand, while growing more slowly (1.4%/yr), is driven by summer peaks and will become even more so, increasing the peak-to-average load ratio and requiring quick-starting turbines and engines to match consumption patterns. Here is where micropower would best match the capacity requirements of utilities and other power providers.

A total of 356,000 megawatts of new generating capacity will be needed by 2025 in the United States, with about half coming online between 2016 and 2025. Of this new capacity, nearly 62% is projected to come from natural gas-fueled turbines or micropower plants.

Micropower's contribution to meeting peak demand could benefit other customers, not just the small generator's owner, by trimming electricity prices. During peak hours, says the U.S. Congressional Budget Office (CBO), "relatively small reductions in demand for utility-supplied power (if the owners of distributed generators produce additional electricity for their own use) or increases in the utilities' supply (if the owners produce additional electricity for sale to the utilities) will reduce wholesale prices considerably."[52]

Micropower's advantages to electric utilities

Micropower promises to help relieve some of the problems facing electric utilities today—peak electricity consumption, competition, and growing demand for more reliable, higher quality power. Also, new technologies are affording utilities better control over unpredictable amounts of electricity streaming in and out of their grid from external sources like fuel cells and other micropower units. This interconnection issue has been a long-standing practical barrier to the acceptance of micropower (see *Barriers and Challenges* in this chapter).

With advances like these, micropower can help the utility meet its peak demand by serving major customers or supplementing the utility's local distribution feeder lines in areas where the grid is highly congested. Computer control systems, operating over telephone lines or the Internet, would allow the utility to use microgenerators as dispatchable resources, supplying electricity as needed.

This is one of the clearest benefits of micropower to the electric industry—it can improve grid reliability by relieving congestion and reducing peak loads. Congestion "drives up prices, sets the stage for cascading outages, and wastes tremendous amounts of energy, as power looping through the system bleeds off as heat," said Steve Silberman in *Wired*.[53]

Utilities are already making deals with customers to share the output of a microgenerator, where the unit's peak capacity is fed into the grid while the user enjoys lower electric rates and a source of emergency power, as well as by-product heat if needed. The utility or power provider also gets ancillary benefits including voltage support and stability, contingency reserves, and quick startup capability.

Just as important to utilities, however, is the desire to defer or avoid construction of long-distance transmission lines and costly upgrades of existing transmission and distribution networks. According to CBO, "Wider adoption of distributed generation can in some cases obviate the need for new transmission capacity." These days, it can cost $365 to $1,100 per kilowatt to run a six-mile power line to a three-megawatt customer,[54] which exceeds the cost of installing many micropower systems and helps make fuel cells competitive.

Also, micropower capacity allows utilities to optimize their existing grid assets—for example, by freeing up their transmission lines for carrying other providers' power—and it offers a cost-effective alternative to building new power lines and transformers, said CBO.

Another advantage of micropower systems is their modular nature. Traditionally, utilities have constructed large, central power plants sized to meet future demand, which is estimated and uncertain. Micropower systems are much more flexible, allowing capacity to be added in small increments that are closely matched with more predictable demand in the near term.

Compared with building conventional plants or transmission lines, siting and permitting are easier, which shortens lead times, and investment is far lower. This limits the utility's capital exposure and risk, letting capital flow toward more productive investments or assets. Micropower systems also have higher energy conversion efficiencies than central plants, yielding more output per dollar of investment.

Additional benefits to utilities and power providers include the potential to sell or trade emission credits and to exploit new business opportunities. Emissions trading is a market-based solution to environmental issues, as opposed to traditional command-and-control schemes. In the United States, the sulfur dioxide trading program has successfully reduced emissions by millions of tons since 1990, and NO_x emissions in the Northeast have been reduced by cap-and-trade programs. The Chicago Climate Exchange offers CO_2 emissions trading and expects this part of its business to expand in the future.

Utilities, other power providers, and energy service companies are participating in hydrogen fuel cell demonstrations. So are many of the world's biggest manufacturers, who see a profitable market in fuel cells, hydrogen production units, storage tanks, and other equipment needed for hydrogen use (see chapter 4).

Government Support For Hydrogen

As mentioned earlier, the European Union and Japan have programs in place to reduce carbon emissions, while the United States relies on voluntary efforts. All three of these governments are implementing policies and funding R&D that promote the use of hydrogen, as are several other nations worldwide. Many scientists believe that government support is essential to accelerate a global transition to a future hydrogen energy economy—or could be required for it to happen at all.

European Union

The European Union (EU) announced in 2002 a bold plan for Europe to become "the first hydrogen economy superpower of the 21st century." The original group of 15 EU nations (which has since expanded to 25) intends to make the historic transition from the fossil-fuel era into a renewable energy economy powered by hydrogen, said the EU.[55] Europe committed to spend more than $2 billion over five years on a hydrogen road map that would bring together industry, research, and government.

The EU already has some of the most ambitious targets for renewable energy in the world. By 2010, renewable resources would provide 22% of the 15 nations' total electricity consumption. Europe leads the world in wind technology, and wind power already accounts for substantial shares of some countries' electricity. However, of the 15 countries, only four—Germany, Denmark, Spain, and Finland—are on track to reach the renewables target. Overall, the EU-15 could achieve 18–19% renewables by 2010, still a respectable share.[56]

The European Union's research is aimed at reducing the cost of hydrogen production, solving storage problems, and integrating renewable sources into hydrogen fuel production over the longer term. One clear motive stated by the EU is to compete with U.S. hydrogen technology. The EU also expects to reap economic benefits from technologies that help improve the environment.[57]

Europe is also working to open up its electricity and natural gas markets, which have been heavily regulated in the past (Table 2–4). This gradual process is now beginning to take effect.

TABLE 2–4. EUROPEAN ENERGY MARKET RESTRUCTURING. WESTERN COUNTRIES ARE WORKING TOWARD LIBERALIZING THEIR POWER AND GAS MARKETS, A PROCESS THAT IS ALSO UNDERWAY IN THE U.S. TABLE COURTESY OF EIA.

	Electricity		Natural Gas	
	Market opening	100% in/by	Market opening	100% in/by
Austria	100%	2001	100%	2003
Belgium	52%	2003/7	59%	2003/6
Denmark	100%	2003	35%	2004
Finland	100%	1997	*	
France	37%	2007	20%	2007
Germany	100%	1999	100%	2000
Greece	34%	2007	*	
Ireland	56%	2005	82%	2005
Italy	70%	2007	100%	2003
Luxembourg	57%	2007	72%	2007
Netherlands	63%	2003	60%	2003
Portugal	45%	2004	*	
Spain	100%	2003	100%	2003
Sweden	100%	1998	47%	2006
United Kingdom	100%	1998	100%	1998
Accession Countries				
Cyprus	-		-	
Czech Republic	30%		0%	
Estonia	10%		80%	
Hungary	30-35%		0%	
Latvia	11%		0%	
Lithuania	26%		90%	
Malta	-		-	
Poland	51%		34%	
Slovakia	41%		33%	
Slovenia	64%		50%	

* Derogation
Source: European Commission, "Second benchmarking report on the implementation of the internal electricity and gas markets", (4/7/2003)

One of Europe's most visible programs is called Clean Urban Transport for Europe (CUTE), which is demonstrating 30 hydrogen fuel cell-powered public buses, three in each of ten cities (see chapter 4). European business leaders have also created the HyNet group, which is working on the Hydrogen Energy Roadmap for Europe. Participants include BP and Shell, BMW, industrial gas supplier Linde, and several utilities. The group provides input to the EU's research programs.

Japan

Japan's lack of oil and natural gas resources has forced it to take an aggressive stance on energy research. As early as 1974, Japan started Project Sunshine, which included a hydrogen budget reportedly in excess of $3 billion through 2000. In the early 1990s, Japan's hydrogen research morphed into a 30-year plan with funding of another $2 billion.

Like other industrial nations, Japan wants to reduce air pollution and improve energy efficiency. But another important goal of its hydrogen research is to create new industries and jobs, because Japan's economy has been languishing since the early 1990s.

All along, Japan's industry has been a global leader in developing fuel cell technology. Japanese utilities are mainly receptive to micropower, and Japan, with some of the world's biggest automakers in its corporate ranks, has also led implementation of hydrogen fuel cells in vehicles and development of fueling infrastructure (Fig. 2–15).

Year 2000	2001	2002	2003
Honda September FCX-V3 (with Ballard stack) Nissan October XTERRA FCV Matsuda *Announced that* *Matsuda would* *participate in the* *project led by* *DaimlerChrysler and* *Nisseki Mitsubishi*	Toyota January *Agreed with GM on clean HC and* *fuel development* Matsuda February Premacy CF-EV Honda February FCX-V3 (with Honda stack) Toyota February FCHV-3 Toyota June FCHV-4 FCHV-BUS1 Honda July FCX-V3 (with Ballard stack) Honda September FCX-V4 (with Ballard stack) Toyota October FCHV-5 Daihatsu October MOVE FCV-K-II	Honda March FCX-V4 (with Ballard stack) Toyota September FCHV-BUS2 Honda October FCX Toyota November FCHV Nissan November X-TRAIL FCV	Daihatsu January MOVE FCV-K-II

High-pressure hydrogen FCV   Methanol reforming FCV   Gasoline reforming FCV

Italic letters: Introduced vehicles   Approved by Minister

FIG. 2–15. JAPAN'S FUEL CELL VEHICLE DEVELOPMENTS. JAPANESE AUTOMAKERS HAVE MOVED VIGOROUSLY INTO THE HYDROGEN FUEL CELL VEHICLE MARKET. ARTWORK COURTESY OF JHFC DEMONSTRATION PROJECT.

One of Japan's main avenues of hydrogen R&D funding is World Energy Network (We-Net), an international program focused on renewable energy with emphasis on hydrogen. As a result of We-Net projects, Japan installed its first two hydrogen vehicle fueling stations in 2002. We-Net entered its third phase in 2003, concentrating on safe application of hydrogen in fuel cells using PEM technology.[58]

Much of Japan's research is directed by the Ministry of Economy, Trade and Industry (METI), which operates the Japan Hydrogen & Fuel Cell (JHFC) demonstration project that began in 2002 (see chapter 4). The JHFC project is Japan's first extensive research that involves actual operation of fuel cell vehicles for demonstration purposes.

Partners in the JHFC program include many of the world's major automakers, as well as Japan's Nippon Oil, Shell, Japanese utilities and gas and equipment suppliers, and even some steel and water companies. METI has specific, ambitious goals for introducing fuel cell vehicles in Japan—50,000 by 2010 and about 5 million by 2020 (Fig. 2–16).[59]

2000-2005	2005-2010	2010~
Infrastructure Organization/Technology Demonstration Stage	Introduction stage	Dissemination stage
• Technology development strategy planning • Institutional–infrastructure organization • Demonstration tests • Fuel quality standard establishment	• Fuel supply system organization • Introduction into public institutions and related enterprises • Second phase of FC technology development strategy	• Fuel supply system organization • Cost reduction • Promotion of adoption by the general population
	Goal of delivering 50,000 fuel cell vehicles by 2010	**Goal of delivering 5,000,000 fuel cell vehicles by 2020**

Fig. 2–16. Japan's fuel cell vehicle goals. The Japanese government has ambitious plans to introduce millions of fuel cell vehicles by 2020. Artwork courtesy of JHFC demonstration project.

Japan's METI and the U.S. DOE agreed in 2004 to pursue pre-competitive research and development of fuel cell and hydrogen technologies. Both countries are members of the International Partnership for the Hydrogen Economy (IPHE), which supports collaborative efforts to deploy hydrogen technologies, common codes and standards for hydrogen fuel utilization, and the sharing of information necessary to develop hydrogen fueling infrastructure.

United States

America's energy policy is a subject of global debate, since the United States consumes such a large share of the world's resources. Historically, policymakers have shied away from taxing gasoline and other transportation fuels, despite heavy fuel taxes in other industrial nations. Suggesting a boost in gasoline taxes has resulted in more than one political downfall, yet taxes are one of the government's primary tools in implementing energy policies that encourage conservation and mitigate air pollution.

As former Energy Secretary James Schlesinger recalls, "An energy tax is surely the best approach—but I've still got the black-and-blue marks from the times I have proposed similar measures in the past! When energy prices are falling, it becomes very difficult to persuade the general public to accept long-term measures."[60]

As an alternative to taxes, the federal government has emphasized R&D funding, and hydrogen has become a highlight of America's energy research. DOE planned to invest a total of $1.7 billion in hydrogen-related R&D through 2010. This investment is explained in DOE's Hydrogen Posture Plan, which outlines the nation's hydrogen energy research in accordance with the earlier National Hydrogen Energy Roadmap.

In 2004 DOE announced $350 million in funding committed to hydrogen research projects. The projects involve 30 lead organizations and over 100 partners. Recipients include academia, industry, and DOE national laboratories. The multi-year funding also supports DOE's FreedomCAR program and Hydrogen Fuel Initiative.

FreedomCAR is a government-industry research partnership that envisions a hydrogen-powered transportation system. The "CAR" stands for Cooperative Automotive Research among DOE, the U.S. Council for Automotive Research (a partnership of DaimlerChrysler, Ford, and General Motors), and the energy industry.

FreedomCAR focuses government support on fundamental, high-risk research that applies to multiple passenger-vehicle models and emphasizes the development of fuel cells and hydrogen infrastructure technologies. FreedomCAR builds upon DOE's earlier Partnership for a New Generation of Vehicles, but shifts the emphasis toward hydrogen fuel.

America's fuel cell research projects also address critical cost and durability issues. The DOE share is $13 million over three years, plus about $10 million in private cost-sharing. In 2004 DOE awarded $75 million for hydrogen fuel research.

Three Centers of Excellence will focus on hydrogen storage research, led by the National Renewable Energy Laboratory, Los Alamos National Laboratory, and Sandia National Laboratories (see chapter 6). Each center will work with several university and industry partners, as recommended by the National Academy of Sciences.

The U.S. hydrogen plan specifies several key milestones:

- Onboard hydrogen storage systems enabling a 300-mile (500-km) driving range

- Hydrogen production from natural gas or liquid fuels at a price equivalent to $1.50 per gallon of gasoline at the pump (untaxed, without carbon sequestration, at 5,000 psi)

- PEM automotive fuel cells that cost $30 to $45 per kilowatt and deliver 5,000 hours of service (equivalent to the vehicle's service life)

- Zero-emission coal plants that produce hydrogen and power with carbon capture and sequestration at 80 cents per gallon of gasoline equivalent (gge) at the plant gate ($1.80/gge delivered)

- Hydrogen production from wind-based electrolysis approaching $2.00/gge (untaxed, using wind electricity at 4 cents per kilowatt-hour, and delivered at 5,000 psi)

- Hydrogen fuel delivery technologies that cost $1.00/gge.[61]

In addition to hydrogen vehicle programs, the United States has formulated a nationwide standard for connecting stationary fuel cells and other micropower plants to the utility grid, which should make them easier to site and build. DOE worked with the power industry to develop the new rules, which were published in 2003 by the Institute of Electrical and Electronics Engineers, Inc. (IEEE). Individual states are in the process of adopting the standard (see next section).

U.S. state programs

Of the 50 U.S. states, 23 offer financial incentives for fuel cell technologies. These incentives include a mix of grants, tax credits, and tax exemptions in Michigan; grants and low-cost loans in Ohio; and grants from Illinois's Clean Energy Community Foundation.[62]

In addition, 23 states are considering adoption of the IEEE's national grid interconnection standards, which should promote the installation of micropower systems (see *Barriers and Challenges* in this chapter), and 15 states are instituting or considering net metering rules, which would allow micropower units to sell electricity back to the grid.

Also, restructuring of the U.S. electric utility industry is proceeding in fits and starts at state levels, which could open the door to high-efficiency micropower sources that can generate electricity at competitive costs. According to EIA, restructuring was active in 18 states, delayed in 5 states, suspended in one (California), and inactive in 27, as of 2003.

Numerous states are initiating hydrogen programs, including New York, Ohio, and Florida. A few of the early initiatives are described here.

California. California leads the states in hydrogen fuel cell vehicle and infrastructure initiatives, primarily through the California Hydrogen Highway program and the California Fuel Cell Partnership. Many of the state's demonstration projects center around Los Angeles, its largest and most polluted city, and Sacramento, its capital (see chapter 4).

State regulations require that by 2008, 10% of new cars sold in California must be zero-emission vehicles, although enforcement of this mandate has slipped in the past. The state tightened its automotive regulations again in 2004, approving America's first limit on CO_2 and other greenhouse gas emissions. However, the new rules, which would take effect in 2009–16, were challenged in court.

Illinois. A similar Hydrogen Highway project is starting up in Illinois, where state, industry, and research partners are funding the first two hydrogen fueling stations, of a total five that are planned. One station was being built at Gas Technology Institute in Des Plaines, and the second was scheduled for construction to start in 2005 at a regional airport in Rockford. This station will be powered in part by solar and wind energy.

Illinois's Hydrogen Highway will stretch along Interstate-90 for about 120 miles (190 km) from Indiana to Wisconsin, with the Chicago area as its geographic center. The five stations would offer hydrogen vehicle fueling every 20 to 25 miles (30 to 40 km). Cost estimates are $600,000–$2 million per station, for a total of $10 million or less. The project is supported by the Illinois $2H_2$ Partnership, a public-private group established by the Illinois Coalition.[63]

Hawaii. As an island where oil imports are costly, Hawaii has been researching hydrogen for almost 20 years and recently boosted its requirements for utilities to use renewable energy sources. By 2015, 15% of utility power sales must draw upon renewables and by 2020, 20%. The state offers a 100% tax credit over five years for high-technology business investments, including fuel cells and renewable energy.

Hawaii hopes to build a Hydrogen Power Park with DOE funding, and the Hawaii Natural Energy Institute on the state university campus opened a fuel cell test facility in 2003. The state is involved in several fuel cell vehicle and micropower development and demonstration projects.

Hawaii also has one of the world's most important atmospheric CO_2 monitoring posts, operated by the U.S. National Oceanic and Atmospheric Administration at the Mauna Loa Observatory. Its data represent the longest continuous record of atmospheric CO_2 concentrations available in the world.

Canada

Like California and Illinois, Canada is planning a Hydrogen Highway in British Columbia along an 80-mile (130-km) corridor from Vancouver to Whistler, with an extension to Victoria. Seven fueling stations are planned, with full implementation targeted for the 2010 Olympics in the two terminal cities.

The Canadian federal government's Natural Resources Canada department is funding the $1.1 million Hydrogen Highway project through the Canadian Transportation Fuel Cell Alliance. Work was underway to supply vehicles and fueling station components.

As a follow-up to the climate change plan issued in 2002, Canada also initiated the h2 Early Adopters (h2EA) program, a government/industry effort to demonstrate hydrogen technologies that further the transition toward a hydrogen economy. Concepts include hydrogen highways and villages (see chapter 4). In addition, the Technology Partnerships Canada program supports individual companies in pre-commercial development projects.

Australia

According to the National Hydrogen Study, all of Australia's buses and 10–20% of other vehicles might run on hydrogen by 2030, and fuel cells could account for up to 5% of commercial and residential micropower.

The province of New South Wales is home to an Energy Centre, which opened in 2003 at the Commonwealth Scientific and Industrial Research Organization in Newcastle. The Centre's building is designed for the future installation of four 25-kilowatt PEM fuel cells.

Also, under Australia's National Research Flagships initiative, the Centre will host a program called Energy Transformed, with a mission to develop low-emission technologies that lead to the use of hydrogen as an energy carrier throughout Australia's economy.

China

China and other Asian countries are stepping up their hydrogen research programs. Some fuel cell vehicles already are operating in China, including a DaimlerChrysler Citaro bus and a bus built by Tsinghua University. China hopes to introduce several fuel cell buses during the 2008 Olympics in Beijing (which China calls the Green Olympics) and the 2010 World Expo in Shanghai. Numerous fuel cell passenger cars and two-wheel vehicles are also being test-driven.

China's research institutions focus generally on PEM fuel cell technology in transportation applications. Among the leaders are the Pan Asia Technical Automotive Center and China's largest auto company, Shanghai Auto Industry Corp., which plans to test a fleet of fuel cell Volkswagen sedans in 2005. General Motors has met with China's leaders to suggest building a hydrogen infrastructure in tandem with the country's expanding gasoline network.[64]

Also, China has substantial installed capacity of micropower systems in industrial sizes. Because grid electricity is notoriously unreliable, up to 10% of the country's total energy supply comes from cogeneration plants, which furnish off-grid power to petroleum refineries, pulp and paper factories, chemical facilities, and iron and steel mills.[65]

Iceland

Iceland decided in 1999 to transform itself into the world's first hydrogen economy. Iceland enjoys abundant hydropower and geothermal energy (underground sources of heat), which are environmentally benign. These power sources are capable of generating more electricity than Iceland consumes—the only country in the world in that position.

Yet Iceland depends on costly oil imports to supply its cars, buses, and ships with gasoline and diesel fuel. The island's transportation and fishing sectors are creating air pollution in a land known for its scenic vistas and pristine water.

Iceland's plan is to use its cheap, clean power to produce hydrogen by electrolysis and ultimately to eliminate petroleum products and use hydrogen fuel cells for transportation. All of the island's 180,000 vehicles and its fleet of more than 2,000 boats will be switched from traditional fuels within 30 to 50 years.

Iceland's status as a developed country, with standards and transport systems similar to other industrial nations, make it an ideal laboratory for this kind of test. Previously, Iceland had shifted all of its space heating equipment from oil to geothermal heat during 1940 to 1975.

Iceland's people support the hydrogen economy, and 93% view fuel cells positively as a replacement for fossil fuel systems. Local enthusiasts hope that the country might eventually export hydrogen to Europe and become renowned globally as the Saudi Arabia of hydrogen.

A joint venture called Iceland New Energy Ltd. was formed to develop hydrogen production technology and fuel cell applications. Demonstration of a hydrogen fueling station and three urban fuel cell buses is under way (see chapter 4).

Norway

A smaller scale project was taking place in Norway's smallest town, the island of Utsira, where energy company Norsk Hydro built the world's first combined hydrogen-wind power plant (see chapter 5). The electrical output from one of two wind turbines is used to make hydrogen, which is stored in a tank and reconverted to electricity on demand by a fuel cell. The second wind turbine sends power into the grid, making the island a net exporter of electricity.

Barriers and Challenges

The vision of a hydrogen economy emerging throughout the world, in industrial and developing nations alike, depends on overcoming many obstacles—chiefly economic and technological, but also regulatory and institutional. Some of these barriers discourage micropower plants, while others threaten the takeoff of fuel cell vehicles, but many are common to both of these hydrogen markets.

Distribution costs

Cost is the biggest challenge—primarily the cost of delivering hydrogen and the complex logistics involved. Distributing hydrogen to isolated vehicle fueling locations would be especially expensive in the early years of a transition when demand for the fuel is widely dispersed, said the U.S. National Research Council (NRC) in its report, *The Hydrogen Economy: Opportunities, Costs, Barriers, and R&D Needs*.[66]

Hydrogen can be transported fairly cheaply by pipeline, but building a new pipeline network for hydrogen is prohibitively expensive in the short term, though it might be a long-term solution in certain strategic areas. Eventually, if hydrogen pipelines are built, the costs of the mature system would be spread over many years, as they have been for today's natural gas network, NRC noted.

For the next few years, compressed or liquefied hydrogen can be trucked to fueling stations. This is too expensive a method to use for very long. According to Sperling and Ogden, "Astoundingly, delivering hydrogen from large plants to dispersed small hydrogen users is now roughly five times more expensive than producing the hydrogen. Even for major fossil fuel-based hydrogen production facilities under study, distribution and delivery costs are estimated to be equal to production costs."[67]

When more cars are on the road, small hydrogen production systems can be installed for onsite reforming of natural gas or other fuels. These distributed fueling appliances are already being developed by major manufacturers. Electrolysis-based generators using water as a hydrogen source could also serve for onsite production as they become more economical to operate (see chapter 7). Onsite hydrogen production, said the NRC report, "avoids many of the substantial infrastructure barriers faced by centralized generation."

However, the NRC report notes that the relationship between hydrogen production, delivery, and dispensing is very complex, even for regional infrastructures to evolve in advance of an integrated national system. "Codes and standards for infrastructure development could be a significant barrier if not established well ahead of the hydrogen market."

Fuel cell technology and onboard hydrogen storage

According to the NRC, the service life of today's fuel cells is much too short, and fuel cell costs are 10 to 20 times too high for market success. In transportation applications, fuel cell systems including onboard storage would have to cost less than $100 per kilowatt ($50 per kilowatt or less in light-duty vehicles) before they would become a plausible commercial option. Otherwise, "fuel cells will not propel the hydrogen economy without some regulatory mandate or incentive."

Fuel cells are also criticized by one of their developers, Trevor Hicks, who worked on the technology at NASA. Citing their poor stack designs and unpredictable life-cycles, he calls upon scientists to make fuel cells more robust. To satisfy consumers, fuel cells in vehicles would have to last about 5,000 hours, which would translate to 150,000 to 200,000 miles.[68]

Also, vehicles will have to store enough hydrogen, without affecting cargo or passenger space, to travel more than 300 miles (500 km), which is one of the U.S. DOE's key goals. As the NRC points out, no one has yet come up with an onboard storage system with an energy density approaching that of gasoline tanks, so fuel cell vehicle range is still much too short (see chapter 6).

Hydrogen production costs

One of the greatest things about hydrogen is that it can be produced from water or from just about any fuel on the planet. But how to do this cost-effectively is a scientific challenge that many ingenious brains are working on right now.

"The laws of thermodynamics are not in our favor," explains Ross Dueber in *Power Engineering*. "It takes more energy to convert compounds like water and methane into molecular hydrogen than the energy we get out when we subsequently use it as a fuel. It doesn't matter if we start with water and use electrolysis or begin with natural gas and perform steam reformation, the end result is the same—less energy out than we put in."[69]

But that's not the whole story. Chip Schroeder explains in RE *Weekly News* that PEM electrolyzers, which are used to produce oxygen in submarines and spacecraft, are similar to a PEM fuel cell running in reverse. From a net energy perspective, the hydrogen end product contains less energy than the electrical power used to produce it.[70]

However, this tradeoff can make economic sense in certain circumstances, he says—for example, small electrolytic generators can be sited anywhere to take advantage of our existing infrastructures for electricity and water, thus minimizing transport and distribution costs. Also, the NRC report points out, as fuel cell costs come down, so will the cost of electrolyzers. At an electrolyzer cost of $125 per kilowatt, the cost of hydrogen would be dominated by the cost of the electricity, not by the cost of the electrolyzer (see chapter 5).

If the electricity used to make hydrogen comes from cheap sources (coal or nuclear) and the hydrogen displaces expensive fuels such as gasoline, says Schroeder, "we have effectively transformed coal or nuclear resources into transport fuel. In such practical applications, the economic value added overwhelms the net energy loss."

However, this pathway to a hydrogen economy is denounced by environmental advocates and mainstream scientists who cite concerns with air pollution and nuclear waste. That's where renewable power comes in, which Schroeder calls "an even more compelling justification for electrolysis.... Renewables give us electricity, but not fuel. The only practical way to turn renewably-generated power (wind, solar, hydro, geothermal) into fuel is through electrolysis."

The NRC report pointed out that tremendous progress has been made in reducing the cost of electricity generated from renewable resources— which would in turn reduce the cost of making hydrogen from renewable power—but further breakthroughs are needed to make renewably produced hydrogen competitive.

Generating the electricity, converting it to hydrogen, and converting it back to electricity obviously adds costs and energy losses, and these are particularly significant when hydrogen becomes a commodity transportation fuel, said NRC. While wind power could approach competitive prices, fundamental research will be needed to develop new processes for generating electricity and producing hydrogen from solar energy (see chapter 5).

Capturing carbon

Producing hydrogen from natural gas and other fossil fuels releases carbon dioxide, which would need to be captured and stored (sequestered). Coal especially would produce massive amounts of CO_2 that would need to be stored reliably for hundreds of years, at least.

Hydrogen production from coal or even from natural gas might not gain broad public acceptance without development of carbon sequestration techniques, said the NRC report. This technical challenge will benefit from knowledge gained in sequestering CO_2 from other operations such as power generation, ammonia synthesis, and oil and gas production (see chapter 5).

Already, some of the world's energy giants are working with governments on CO_2 capture, and some environmentalists have decided to keep an open mind on the subject. As Vaitheeswaran concludes, "Sequestration could well act as a stepping-stone to a renewable-based hydrogen economy."[71]

Regulatory and institutional barriers

Many of the economic factors discussed so far could affect the market for hydrogen in stationary micropower applications, but not as drastically as in transportation. Fuel cell costs would have to fall, and their lifetimes would have to be extended for widespread use in stationary markets. But hydrogen for micropower plants would almost certainly be produced onsite, eliminating transport and distribution costs, and hydrogen storage tanks would not have to be as lightweight as on vehicles.

Instead, most of the risks and barriers affecting stationary micropower are regulatory and institutional—that is, they involve utilities' pricing and operational practices and local governments' rules about reliability, safety, cost, and environmental quality.

Interconnection requirements. Historically, the biggest obstacle to micropower has been connecting units with the utility grid. Most customers require interconnection for emergency power (standby or backup power), supplemental electricity, and in some cases for marketing or wheeling microgenerated power. The key to the ultimate market success of small micropower systems is the ability to safely, reliably, and economically connect to the utility grid.

In today's markets, despite deregulation, utilities are still responsible for maintaining the safety and reliability of the grid, so they naturally have concerns about the interconnection of other equipment to the network. It's difficult for utilities to control an unpredictable amount of electricity coming in and out of their grid from external sources.

The assumption that power flows in one direction is deeply embedded in the protective relays of the existing system, and influxes of power from unexpected sources can endanger utility personnel. Also, electricity can't be effectively stored, and electrons do not behave in an orderly fashion, so supply must be orchestrated to meet demand with split-second precision.[72] But interconnection requirements are often duplicative, excessive, and time-consuming, say micropower advocates, and the utility approval process must be standardized and streamlined.

Contractual and technical requirements for installing equipment and safety devices are needed to protect the grid and ensure power quality. However, according to EPRI *Journal*, some local authorities impose their own specific interconnection requirements, while others require case-by-case interconnection studies similar to those for conventional large power plants. In many areas existing rules are applied in an *ad hoc* fashion, and some agencies ignore micropower altogether.[73]

The resulting patchwork of inconsistent interconnection regulations adds cost, complexity, delays, and uncertainty to the process of developing micropower projects, and this barrier has slowed investment in new technologies. This is especially true of smaller systems, where such factors have a disproportionate effect on the project's total cost. Excessive costs and delays can effectively sabotage micropower applications in the smaller size range.

As mentioned above, the United States and its individual states are making progress toward overcoming this barrier. The IEEE standard could be used in federal legislation and rulemaking, in state utility commission deliberations, and by more than 2,500 electric utilities in formulating technical requirements for interconnection agreements, said EPRI *Journal*.

Japan has encouraged micropower cogeneration by removing several regulatory obstacles, including adjustments to fire regulations and onsite staffing levels. However, Japanese micropower operators generally cannot sell excess power to other customers, and grid-protection equipment can account for 10% of the installation's total cost.[74]

Utility tariffs and rates. Another potential showstopper is the utility tariff and rate structure, which can discriminate against micropower. Some utilities charge excessive rates or fees for standby service while others apply exit fees to customers who want to become independent of the utility grid. These surcharges do not reflect the actual cost of the service, nor do they credit micropower's benefits to the grid, such as supplying on-peak electricity.

Other rates and tariffs include stranded-cost recovery charges (also called competitive transition charges), which are usually applied on a temporary basis to allow utilities to adjust to restructuring. Additionally, many utilities are allowed to renegotiate rates if they determine that a customer is intending to implement micropower. Often, the utility will offer a reduced rate to retain the customer and avoid a micropower project (a practice called selective discounting).

Electricity pricing. U.S. utilities are required by federal law to purchase electricity from cogenerating micropower plants at the utility's avoided cost (whatever it would have cost the utility to generate the power). However, utilities can base the rate on their average cost of power rather than their marginal cost—the cost of supplying the last additional unit of electricity, which is much higher. Thus the utility buyback rate is often low enough to discourage a micropower project if its economics depend on selling excess power.

Utility pricing also exacts high rates for peak power, which are called demand charges. Commercial and industrial customers bear the brunt, but power providers could institute time-of-day rates (also called real-time pricing) for residential customers in the future. These price structures reflect how expensive it is for the utility to supply that extra increment of power during peak periods. Wholesale electricity prices also spike on hot summer days and other peaks. Micropower could offer customers relief from peak prices by providing electricity when it's needed most.

Environmental permitting. Local siting restrictions and permitting requirements, including air quality regulations, can impose burdensome approval processes that inhibit micropower projects, especially in smaller sizes because the costs of siting and permitting don't scale down fully with project size.

Also, air quality requirements fail to recognize micropower's lower emissions output compared to conventional, less efficient sources of electricity. Micropower projects should receive credit for avoided or displaced emissions, say proponents. Local and regional building codes, with more than 44,000 jurisdictions, also need to be addressed, said the Congressional Budget Office.

Risk of electric price increases

According to the CBO, micropower promises to encourage efficient investments that improve grid reliability, expand service, and reduce congestion. But these benefits depend on utilities being able to manage a much greater number of power sources. If ratepayer-funded investments are needed to maintain power quality and reliability, retail electricity prices could rise.

Also, poorly designed policies and financial incentives could prompt customers to invest in micropower equipment whose output is more expensive than new central power, said CBO. This outcome could increase the overall cost of electricity to the utilities' remaining customers.

In the United States, battles are being fought at the state level over this very issue, dividing potential micropower operators into opposing groups— large commercial and industrial consumers, who would typically install cogeneration systems producing a megawatt or more of power, and small businesses and homeowners, who might install fuel cells, microturbines, and solar energy systems as small as a few kilowatts.

Safety

Safety is a technical barrier that most hydrogen advocates believe is fairly easy to resolve. The risks of using hydrogen are relatively low compared with gasoline, propane, and other common fuels. Still, safety is an especially important issue in the hydrogen vehicle market (see chapter 6).

Experts differ on hydrogen's safety in a consumer-centered transportation system. For example, an underexplored topic is leakage in enclosed structures such as household garages and commercial buildings. The NRC report recommends physical testing and development of standards in advance of commercial use.

In a future hydrogen economy, we might eventually store and transport hydrogen in ways very similar to how we handle natural gas and propane today. But consumers' perceptions of hydrogen's dangers would need to be alleviated, and codes and standards would have to be further developed.

Outlook For a Transition to a Hydrogen Economy

The hydrogen economy has many serious detractors. Swiss fuel cell proponent Ulf Bossel is one:

> Mankind needs fuel cells not as stepping stones for a Hydrogen Economy, but because of their high efficiency, their environmentally benign behavior and their potentials for small-scale power generation. Mankind needs fuel cells for hydrocarbon fuels, be it with [an] external reformer, or with internal reforming.

*What mankind does not need is the intellectual alliance between fuel
cells and a Hydrogen Economy. The fuel cell industry is threatened
by the promotion of hydrogen by individuals who are neither linked
to the hydrogen market nor involved in the development of fuel cells,
and whose claims are based on visions rather than on hard fact of
science and engineering.*[75]

Bossel and other analysts of the hydrogen economy's physics advocate
methanol, ethanol, or dimethyl ether as preferred energy carriers in a renew-
ables-based economy.

Another, more colorful naysayer is John R. Wilson of TMG/The Management Group, who says, "Hydrogen has been greatly oversold by 'evangelists'
in the USDOE and elsewhere and also by the environmental lobby, including
some very persuasive writers who are adept at choosing half truths to fit their
preconceived conclusions. In short, upon close and objective examination,
we find that The Emperor Hydrogen has no Clothes."[76]

Even NRC envisions that the path toward a hydrogen economy is neither simple nor straightforward, and it presents major hurdles. "The hydrogen system must be cost-competitive, it must be safe and appealing to the
consumer, and it would preferably offer advantages from the perspectives of
energy security and CO_2 emissions. Specifically for the transportation sector,
dramatic progress in the development of fuel cells, storage devices, and distribution systems is especially critical. Widespread success is not certain."

Other observers are more optimistic. Vijay Vaitheeswaran believes that
we have a chance to set the world's energy system on a more sustainable
foundation. "The good news is that, after decades of stagnation, the energy
realm is wide open to new ideas, new technologies, and especially *new ways
of thinking.*"[77]

Jeremy Rifkin, in his book *The Hydrogen Economy*, also envisions an opportunity to seize. "There are rare moments in history when a generation of human beings are given a new gift with which to rearrange their relationship to one another and the world around them. This is such a moment. We are being given the power of the sun. Hydrogen is a promissory note for humanity's future on Earth."[78]

References

1. Sperling, Daniel, and Ogden, Joan, "The Hope for Hydrogen," *Issues in Science and Technology*, Spring 2004, page 86.

2. Vaitheeswaran, Vijay V., *Power to the People: How the Coming Energy Revolution Will Transform an Industry, Change Our Lives, and Maybe Even Save the Planet*. Farrar, Straus and Giroux: New York, 2003. ISBN 0-374-23675-5, page 327.

3. Hoffman, Peter, *Tomorrow's Energy: Hydrogen, Fuel Cells, and the Prospects for a Cleaner Planet*. The MIT Press: Cambridge, MA, 2001. ISBN 0-262-08295-0, page 13.

4. Vaitheeswaran, page 140.

5. Claussen, Eileen, Pew Center on Global Climate Change, Vaitheeswaran, page 138.

6. Vaitheeswaran, page 131.

7. Claussen, Eileen, Pew Center on Global Climate Change, in Vaitheeswaran, page 139.

8. Intergovernmental Panel on Climate Change (United Nations), in Vaitheeswaran, pages 139–140.

9. Wallace, John M., and Christy, John R. , "The Truth About Global Warming," February 4, 2000, www.nas.edu/nrc.

10. McNeill, John, Georgetown University, in Vaitheeswaran, page 129.

11. Intergovernmental Panel on Climate Change (United Nations), in Vaitheeswaran, pages 132, 138.

12. National Academy of Sciences, "Climate Change Science: An Analysis of Some Key Questions," 2001, www.nap.edu.

13. European Environment Agency, *Impacts of Europe's changing climate*, August 18, 2004, www.eea.eu.int.

14. Rocky Mountain Institute, www.rmi.org.

15. United Nations Framework Convention on Climate Change, http://unfccc.int.

16. European Environment Agency, "Domestic measures taken or planned so far are insufficient to meet EU climate emissions targets, projections show," December 2, 2003, http://org.eea.eu.int.

17. "Carbon-emissions trading: A green future," *The Economist*, September 11, 2004, pages 69–70.

18. Vaitheeswaran, pages 122, 156–157, 159

19. Hoffmann, Peter, *Tomorrow's Energy: Hydrogen, Fuel Cells, and the Prospects for a Cleaner Planet.* The MIT Press: Cambridge, MA, 2001. ISBN 0-262-08295-0, page 17.

20. U.S. Energy Information Administration, *International Energy Outlook* 2004, www.eia.doe.gov/oiaf/ieo.

21. Silberman, Steve, "The Energy Web," *Wired*, Issue 9.07, July 2001, www.wired.com.

22. Töpfer, Klaus, United Nations Environment Programme, in Vaitheeswaran, page 173.

23. Rifkin, Jeremy, *The Hydrogen Economy: The Creation of the Worldwide Energy Web and the Redistribution of Power on Earth.* Penguin Group (USA) Inc.: New York, 2002. ISBN 1-58542-193-6, page 239.

24. World Bank, World Development Report 2003, in Vaitheeswaran, pages 170–172.

25. U.S. Energy Information Administration, *Annual Energy Outlook* 2004 *with Projections to* 2025, www.eia.doe.gov/oiaf/aeo.

26. Rocky Mountain Institute, www.rmi.org.

27. Lord Browne, "Climate Change," Institutional Investors Group, Bishopsgate, London, November 26, 2003, www.bp.com.

28. Ausubel, Jesse, Rockefeller University, in Vaitheeswaran, page 187.

29. Rocky Mountain Institute, www.rmi.org.

30. Vaitheeswaran, page 12.

31. Ibid.

32. Nering, Evar, Arizona State University, in Vaitheeswaran, page 11.

33. Duncan, Richard C., "Big jump in ultimate recovery would ease, not reverse, postpeak production decline," *Oil & Gas Journal*, July 19, 2004, page 18.

34. Samsam Bakhtiari, A. M., "Middle East oil production to peak within next decade," *Oil & Gas Journal*, July 7, 2003, pages 20–28.

35. Samsam Bakhtiari, A. M., "World oil production capacity model suggests output peak by 2006-07," *Oil & Gas Journal*, April 26, 2004, pages 18–20.

36. Imam, Asher, Startzman, Richard A., and Barrufet, Maria A., "Multicyclic Hubbert model shows global conventional gas output peaking in 2019," *Oil & Gas Journal*, August 16, 2004, pages 20–28.

37. al-Jubeir, Adel, in *The Wall Street Journal*, May 27, 2004.

38. Maugeri, Leonardo, "Time to debunk mythical links between oil and politics," *Oil & Gas Journal*, December 15, 2003, page 22.

39. Jenkins, Holman W. Jr., "Nothing to Say? Talk About 'Energy Independence',"
    The Wall Street Journal, June 2, 2004, page A15.

40. Rifkin, page 36.

41. Rocky Mountain Institute, www.rmi.org.

42. Vaitheeswaran, page 26.

43. Sperling, Daniel, and Ogden, Joan, page 85.

44. Chambers, Ann. *Distributed Generation: A Nontechnical Guide.* Tulsa, OK: PennWell,
    2001, page 22.

45. Congressional Budget Office, "Prospects for Distributed Electricity Generation,"
    September 2003, www.cbo.gov.

46. Vaitheeswaran, page 34.

47. Edison Electric Institute, "Industry Statistics," May 2003, www.eei.org.

48. Parks, Bill, U.S. Department of Energy, "Transforming the Grid to Revolutionize
    Electric Power in North America." Paper presented at the Edison Electric Institute
    Fall 2003 Transmission, Distribution and Metering Conference, October 13, 2003.

49. "Blackout Cost New York $36m an Hour," Guardian Newspapers Ltd.,
    August 19, 2003.

50. Samotyj, Marek, "The Cost of Power Disturbances to Industrial & Digital
    Economy Companies." Report by Primen for Electric Power Research Institute,
    Consortium for an Electric Infrastructure to Support a Digital Society (CEIDS),
    June 2001.

51. Business Communications Co., "The Power Quality Equipment and Services
    Market: A Growing 21st Century Business," November 2001.

52. Congressional Budget Office, "Prospects for Distributed Electricity Generation,"
    September 2003, www.cbo.gov.

53. Silberman, Steve, "The Energy Web," *Wired*, Issue 9.07, July 2001, www.wired.com.

54. Chambers, page 22.

55. European Union, "EU announces plan for Hydrogen Economy,"
    October 24, 2002.

56. European Commission, "Renewable energy: Commission calls for a
    stronger commitment of Member States to achieve the 2010 targets,"
    May 26, 2004, http://europa.eu.int.

57. European Commission, Energy Research Program, http://europa.eu.int.

58. World Energy Network (We-Net), www.enaa.or.jp/WE-NET.

59. Japan Hydrogen & Fuel Cell Demonstration Project, www.jhfc.jp.

60. Schlesinger, James, in Vaitheeswaran, page 324.

61. U.S. Department of Energy, Office of Energy Efficiency and Renewable Energy, DOE *Hydrogen Posture Plan*, February 2004, www.eere.energy.gov/hydrogenandfuelcells/pdfs/hydrogen_posture_plan.pdf

62. Database of State Incentives for Renewable Energy (DSIRE), www.dsireusa.org.

63. Illinois 2H$_2$ Partnership, "The Hydrogen Highway: Illinois' Path to a Sustainable Economy and Environment," March 2004, www.ilcoalition.org.

64. Breakthrough Technologies Institute, *Fuel Cell Vehicle World Survey* 2003, February 2004, www.fuelcells.org.

65. Silberman, Steve, "Girding Up For the Power Grid," *Wired*, June 14, 2001, www.wired.com.

66. National Research Council, *The Hydrogen Economy: Opportunities, Costs, Barriers, and R&D Needs.* The National Academies Press, 2004 (394 pages). ISBN 0-309-09163-2, www.nap.edu.

67. Sperling, Daniel, and Ogden, Joan, page 85.

68. Hicks, Trevor, in Silverstein, Ken, "Fuel Cells Get Big Lift," Issue Alert, May 9, 2003, www.utilipoint.com.

69. Dueber, Ross, in Blankinship, Steve, "When Will the Hydrogen Future Arrive?", *Power Engineering*, January 2003, http://pe.pennwellnet.com.

70. Schroeder, Chip, "Hydrogen from Electrolysis," RE *Weekly News*, June 21, 2004, www.solaraccess.com.

71. Vaitheeswaran, page 249.

72. Silberman, Steve, "Girding Up For the Power Grid," *Wired*, June 14, 2001, www.wired.com.

73. "IEEE Distributed Energy Interconnection Standard Published," EPRI *Journal*, August 20, 2003, www.epri.com.

74. International Energy Agency, "Distributed-Generation Issues in Japan, the US, the Netherlands, and the UK," www.iea.org.

75. Bossel, Ulf, "The Physics of the Hydrogen Economy," *European Fuel Cell News*, Volume 10, Number 2, July 2003, page 16.

76. Wilson, John R., and Burgh, Griffin, "The Hydrogen Report: An Examination of the Role of Hydrogen In Achieving U.S. Energy Independence," TMG/The Management Group, July 2003, www.tmgtech.com.

77. Vaitheeswaran, pages 320–321.

78. Rifkin, pages 254–255.

How would Hydrogen be Used?

In future energy systems, hydrogen would serve as an energy carrier like electricity does today. The energy used to produce hydrogen would be delivered to customers as electric power and heat. Much of our hydrogen supplies would be used in fuel cells, which generate power on demand. Microturbines, engines, and larger turbines could also burn hydrogen or hydrogen-natural gas fuel mixtures.

Hydrogen's biggest advantage over traditional hydrocarbon fuels is that hydrogen can be pollution-free. When combined with oxygen in a fuel cell, hydrogen produces only electricity and water vapor. However, burning hydrogen with air can produce some emissions, and many of hydrogen's properties make it difficult to store and transport cost-effectively, though less so than electricity.

In a future hydrogen economy, fuel cells could provide power in three broad markets: automotive, stationary, and portable. In automotive applications, fuel cells and electric motors would replace engines in cars, trucks, and buses, as well as in ships and possibly trains and small planes. Hydrogen's primary advantage in these transportation uses is that tailpipe exhaust is reduced or eliminated. Fueling stations would dispense hydrogen for vehicles, and equipment would become available for home fueling.

In stationary micropower markets, fuel cells could provide electricity for homes, businesses, industry, and utilities, complementing conventional power. Hydrogen fuel cells could replace batteries in uninterruptible power supply systems, for example, or generate electricity during peak demand periods or emergencies. Many stationary fuel cells would incorporate reformers to convert natural gas into hydrogen. In portable applications, fuel cells would replace batteries in laptops, cellphones, and other small devices. In these micropower and portable markets, fuel cells have the advantage never losing their charge like batteries do, as long as hydrogen is supplied.

In terms of market maturity, portable applications of fuel cells are closest to commercial use. Stationary fuel cell micropower plants are already on the market but are too expensive for widespread application. Niche markets include customers who require high power quality and reliability. Automotive fuel cells are being tested in a variety of vehicles worldwide and could begin to enter commercial markets around 2015 to 2020.

Fuel cells are classified by their type of electrolyte: phosphoric acid, molten carbonate, solid oxide, and alkaline, as well as polymer electrolyte membrane (or proton exchange membrane), which is called PEM. Most automotive and portable applications use PEM fuel cells, while phosphoric acid and solid oxide systems are better suited for stationary micropower. Molten carbonate fuel cells could supply utility and industrial power, while alkaline fuel cells would continue to be used in aerospace applications.

These topics are discussed in Chapters 3 and 4. Chapters 5, 6, and 7 discuss hydrogen's production, transport, storage, and safety.

Fundamentals and **Background**

Hydrogen is the source of the sun's heat and light, which is radiated to the earth in enormous amounts. Hydrogen-fueled reactions also supply most of the energy given off by other stars. When we use gasoline or natural gas, we are actually drawing upon the sun's energy, which was stored ages ago by ancient plants.

Hydrogen is a distinctive element in many ways. Its atomic structure is the smallest, lightest, and simplest of all the elements. It consists of just a single proton and a single electron (subatomic particles with electrical charges). A string of more than 40 million hydrogen atoms would be less than half an inch long. Yet these minuscule hydrogen atoms make up about 90% of all matter in our universe, and the human body consists primarily of hydrogen.

Our Supply of Hydrogen

On earth, however, hydrogen is not found free in nature. Hydrogen combines with all other elements except inert gases, so it creates more compounds than any other element. Virtually none of the earth's hydrogen is in a form that we can readily use in fuel cells or other energy applications.

Instead, hydrogen is nearly always bonded with other elements—with oxygen in water, for example, or with carbon in hydrocarbon fuels such as natural gas, oil, and coal. Almost all organic compounds, which by definition contain carbon, also contain hydrogen.

Hydrogen can be produced by unlocking the chemical bonds in water, hydrocarbon fuels, and other compounds. But these methods consume energy, because the chemical bonds in these substances are strong and stable. Water is so stable chemically (nonreactive) that it cannot be burned as a fuel.

However, because hydrogen is found in so many different fuels and compounds, it can be produced from any energy source—and it can be obtained from one of our planet's most common substances, water. Producing hydrogen from water also yields oxygen as a valuable by-product.

All of this means that the earth's supply of hydrogen is not a source of energy like oil or gas. Instead, hydrogen is usually considered an energy carrier—a way to transport and store energy that is produced elsewhere, from other sources. In a similar way, electrons serve as an energy carrier today in the form of electric power, which is also produced from a variety of fuels and energy sources.

Hydrogen's Properties

At room temperature, hydrogen is a gas with two atoms per molecule (H_2). Gaseous hydrogen is colorless, odorless, tasteless, and nontoxic. Some of hydrogen's properties make it less convenient to use than today's common fuels, while other properties give hydrogen an advantage over them.

The hydrogen molecule's tiny size makes it more difficult to store than natural gas (primarily methane, or CH_4). Hydrogen can escape from most containers by squeezing through the structure of whatever materials are used to hold it.

Also, hydrogen's light weight means that it takes up a lot of space. Compressing hydrogen gas to fit into a storage tank requires energy, more than is needed to compress natural gas.

Hydrogen turns into a liquid only at extremely cold temperatures (–253°C, or –423°F), in the cryogenic range. Liquefying hydrogen gas also consumes energy, as does liquefaction of natural gas. Oil, gasoline, and other fuels that are liquid at room temperature are easier to store and transport than hydrogen.

But hydrogen contains more energy per unit of weight of any known fuel—about three times the energy of gasoline weighing the same amount and almost seven times that of coal. Hydrogen's energy content is 52,000 Btu per pound (120.7 kilojoules per gram). This factor is one advantage of hydrogen over other fuels.

Hydrogen has a low energy content per unit of volume, however—only about one-third the energy of hydrocarbons taking up the same amount of space, at normal temperatures and pressures. That's why hydrogen has to be compressed or liquefied to store it in useful quantities.

When cooled to a liquid, hydrogen takes up just 1/700th as much space as it does in its gaseous form. That's one reason hydrogen is used as a fuel in rockets and space missions, which require an extremely lightweight, compact, high-energy fuel. Hydrogen can also be stored in compounds called metal hydrides, and possibly in the future in microscopic structures made of carbon (see chapter 6).

Hydrogen is highly flammable in the presence of air. Only a small amount of energy is needed to ignite hydrogen and make it burn. Also, it has a wide range of flammability, meaning that it will burn in low concentrations as well as high ones (4% to 74% by volume in air).

Also, hydrogen can't be detected by people's sense of smell, so a leak of pure hydrogen could be hazardous if it accumulates (see chapter 6). And hydrogen burns with an almost invisible pale-blue flame, making hydrogen fires hard to see. However, because hydrogen is so light and buoyant, it dissipates more rapidly in air than natural gas, reducing the risk of explosions, and it cannot seep into the ground like oil or gasoline.

The Cleanest Fuel

Compared with traditional fossil fuels, hydrogen is unique in that it produces no harmful emissions when it's combined electrochemically with oxygen. The only product is pure water, which returns to the earth's natural cycle. Amidst all the hype about hydrogen, this single fact stands out as its most important advantage over today's fuels: hydrogen can be pollution-free.

In contrast, all hydrocarbon fuels produce carbon dioxide (CO_2) during combustion. Essentially, this is the reverse of photosynthesis, in which plants consume CO_2 in order to grow. When fossil fuel is burned, its carbon combines with oxygen from the air to form CO_2 emissions. Many natural processes—like human respiration—also generate CO_2.

Carbon dioxide is regarded as a primary culprit in the gradual process of global warming. The presence of CO_2, methane, and other greenhouse gases in our atmosphere is responsible for this warming trend, many scientists think (see chapter 2). These gases allow the sun's rays in to warm the earth, but they prevent heat from being radiated back into space.

Fossil fuels also produce other air pollutants when burned in automotive engines and power plants. These include nitrogen oxides (NO_x), which are an ingredient in urban smog, and sulfur oxides (SO_x). Oxides of both nitrogen and sulfur react to form acid rain. Hydrogen also can form NO_x when it's burned with air, which contains about 80% nitrogen.

Other air pollutants created from hydrocarbon fuel consumption are poisons such as carbon monoxide, various toxins, and another smog constituent called VOCs. Hydrogen creates none of these pollutants.

Some hydrocarbon fuels—most notably coal and diesel—also generate smoke and soot emissions (particulates). These microscopic particles can affect human health, especially in vulnerable people who have breathing problems such as asthma and emphysema. Hydrogen produces zero particulate emissions, and those of natural gas are also quite low.

Producing Hydrogen

Hydrogen can be produced from water, hydrocarbon fuels, or other compounds. For more details on hydrogen production methods, see chapter 5.

Electrolysis

A water molecule contains two hydrogen atoms and one oxygen atom (H_2O). Water-splitting, or electrolysis, produces the two gases, hydrogen and oxygen, when an electric current is applied. In the reverse reaction, hydrogen combines with oxygen, producing water and giving off large amounts of heat (an exothermic reaction).

$$2H_2O \rightarrow 2H_2 + O_2$$
$$2H_2 + O_2 \rightarrow 2H_2O$$

During electrolysis, electricity enters a container of water at one end—the cathode, a negatively charged electrode that repels electrons, which carry a negative charge. The water is made conductive by adding an electrolyte, either acidic or basic. The current passes through the water and exits at the other end—the anode, a positively charged electrode that attracts electrons and repels protons, which carry a positive charge. Hydrogen collects at the cathode and oxygen at the anode.

Electrolysis does not require significant amounts of water. The hydrogen extracted from a gallon of water could drive a vehicle about as far as gasoline vehicles travel today on a gallon of gasoline.[1]

In the mirror image of electrolysis, hydrogen is combined with pure oxygen, and nothing but water is formed, along with the heat released by the reaction. The amount of energy produced is the same amount needed to split water, according to the basic laws of thermodynamics. The electricity produced is in the form of direct-current power, rather than alternating-current (see *Fuel Cell Technology* in this chapter).

Today hydrogen can be produced by electrolysis, but it's expensive because it uses a lot of electricity. Commercial water-splitting consumes power and usually uses a catalyst (a substance that accelerates chemical reactions), which is made of costly materials.

However, electrolysis is a proven, efficient process and can be economical in areas where power is very cheap, such as Canada and Norway, which possess vast hydropower resources.[2] Only about 4% of the world's hydrogen supply comes from electrolysis, mostly for use in processes requiring very pure hydrogen, such as electronics manufacturing.

Although the electrical energy needed to split water exceeds the energy contained in the hydrogen product, eventually this electric power could come from wind turbines, solar energy (Fig. 3–1), or other renewable resources.

Fig. 3–1. Solar power facility in California. One of the world's largest solar power facilities, built in the late 1980s in California, generates a total of 150 megawatts. Photo by Kramer Junction Co., courtesy of NREL.

Reforming

The least expensive way to produce hydrogen today is from hydrocarbon fuels, which are "re-formed"—that is, their chemical bonds are broken and the components rearranged into new forms, using a catalyst and heat in the form of steam. This thermal energy strips hydrogen from the carbon and other elements in the fuel.

Nearly all of the hydrogen produced today (96%) comes from steam re-forming of fossil fuels, mostly natural gas (48%). Some is also produced from reforming oil (30%) and coal (18%). All of these hydrocarbon fuels produce carbon dioxide when they are reformed. The carbon in the fuel combines with oxygen to form CO_2.

In the natural gas reforming process, methane reacts with steam at high temperatures, accelerated by a catalyst. This produces a mixture of hydrogen, carbon monoxide, and other gases. Then a shift reaction produces more hydrogen and changes the carbon monoxide to carbon dioxide (see chapter 5).

Other processes

Several advanced techniques are being developed to produce hydrogen in the future. These include gasification of coal and biomass, which produces a gas mixture from which hydrogen can be extracted.

Two new methods being developed would use solar energy to split the water molecule: photobiological techniques exploit algae and bacteria to produce hydrogen in the presence of sunlight, and photoelectrochemical processes split water into hydrogen and oxygen using sunlight and semi-conductor electrodes.

The water molecule can also be split into hydrogen and oxygen using heat and chemicals, at very high temperatures. In the future, these thermochemical methods might be powered by solar or nuclear energy (see chapter 5).

Fuel Cell Technology

Fuel cells are similar to batteries, but instead of generating power from stored chemicals, fuel cells combine hydrogen with oxygen in an efficient electrochemical process that is clean and flameless. Essentially, this is the reverse of electrolysis. Nothing is produced but electric power and pure, distilled water, along with heat that can be recovered and used. Unlike batteries, fuel cells need no recharging; they will operate as long as hydrogen fuel is supplied.

In a fuel cell, the hydrogen molecule's atoms are split into protons and electrons. The fuel cell's electrolyte lets the protons pass through, but it diverts the electrons through external wiring, allowing their energy to be captured as electricity. When the protons and electrons are reunited, the hydrogen molecules combine with oxygen to form water. Catalysts are often used to accelerate the process of breaking up the hydrogen molecule.

To obtain the desired amount of electrical power, individual fuel cells can be connected into stacks. Increasing the stack's number of cells boosts its voltage, while increasing the cells' surface area boosts its current. Like other power generators, fuel cells release heat in varying amounts. When this heat is recovered and used in cogeneration applications, the overall system efficiency increases.

Types of fuel cells

Decades of research have resulted in the evolution of different fuel cell technologies, which fall into a handful of categories based on their type of electrolyte. These include phosphoric acid, molten carbonate, solid oxide, and alkaline cells, as well as PEM. Each fuel cell type has pros and cons in particular applications.

PEM. PEM fuel cells deliver high power density and offer the advantages of low weight and volume, compared to other types. They operate at relatively low temperatures, around 80°–100°C (176°–212°F). Low-temperature operation allows them to start quickly (less warm-up time) and results in less wear on system components, which makes the PEM cell durable. Also, PEM fuel cells respond rapidly to changes in demand for power, and they do not require corrosive fluids as some types do. All of these factors make PEM fuel cells good candidates for vehicles and micropower applications.

PEM technology uses a solid polymer membrane (thin plastic film) and porous carbon electrodes containing a platinum catalyst or other noble metal (Fig. 3–2). The catalyst is expensive and also makes the cell extremely sensitive to poisoning by carbon monoxide (CO), which must be eliminated in the fuel gas along with sulfur and other impurities. Developers are exploring catalysts that are more resistant to CO.

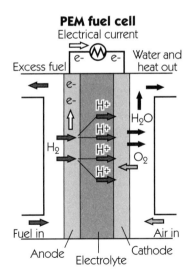

PEM fuel cell

Fig. 3–2. PEM fuel cell schematic. Lightweight, low-temperature PEM fuel cells are a good choice for automotive applications. Artwork courtesy of DOE.

Phosphoric acid. These fuel cells use liquid phosphoric acid as an electrolyte—the acid is contained in a Teflon-bonded silicon carbide matrix—and porous carbon electrodes containing a platinum catalyst. Phosphoric acid fuel cells are used mainly for stationary power generation, but some have been applied in large vehicles such as city buses. The phosphoric acid fuel cell is considered the first generation of modern fuel cells. It is one of the most mature cell types and the first to be used commercially, with hundreds of units currently in operation.

Phosphoric acid fuel cells are less susceptible to CO poisoning than PEM types. They are up to 80% efficient in cogenerating electricity and heat, but only 37% to 55% efficient at producing electricity alone. Phosphoric acid fuel cells also deliver less power than other types, so they are usually large and heavy.

Solid oxide. For industrial facilities, solid oxide fuel cells are a good fit because they generate large amounts of useful high-temperature heat. Solid oxide fuel cells are around 45% to 65% efficient at converting fuel to electricity. In cogeneration applications, fuel use efficiencies could reach 85%.

Solid oxide fuel cells use a hard, non-porous ceramic compound as the electrolyte. They operate at very high temperatures—around 1,000°C (1,800°F), which means they do not require an expensive precious-metal catalyst. High-temperature operation also allows solid oxide cells to reform fuels internally, so no costly reformer is needed, and a variety of fuels can be used. Solid oxide fuel cells are also the most sulfur-resistant type, so they can run on gases made from coal.

The disadvantages of high-temperature operation include slow start-up and requirements for heat shields and highly durable materials.

The development of low-cost materials with high durability at cell operating temperatures is the key technical challenge facing this technology.

Alkaline. Alkaline fuel cells were one of the first fuel cell technologies developed, and they were the first type widely used in the U.S. space program to produce electricity and water onboard spacecraft. These fuel cells use a solution of potassium hydroxide in water as the electrolyte and can use a variety of non-precious metal catalysts, but they require relatively pure hydrogen and oxygen.

High-temperature alkaline fuel cells operate between 100°C and 250°C (212°F and 482°F). However, more recent designs operate at lower temperatures, roughly 23°C to 70°C (70°F to 160°F). Alkaline fuel cells reach efficiencies of 60% in space applications, but are easily poisoned by CO_2, and purification of fuel and oxygen is costly. Also, their operating lifetimes are relatively short.

Molten carbonate. These high-temperature fuel cells (650°C, or 1,200°F and above) are suitable for utility power plants and industrial applications and do not need precious-metal catalysts. They use an electrolyte composed of a molten carbonate salt mixture suspended in a porous, chemically inert ceramic lithium-aluminum oxide matrix.

The efficiencies of molten carbonate fuel cells approach 60%, considerably higher than phosphoric acid types. With cogeneration, they can reach 85%. Like solid oxide types, molten carbonate cells require no external fuel reformer, nor are they prone to poisoning by CO or CO_2. However, their electrolyte is corrosive and cell life is short.

These fuel cell types are summarized in Table 3–1.

TABLE 3–1. FUEL CELL TECHNOLOGIES

Fuel cell type	Temp. range	Efficiency*	Electrolyte	Capacities	Primary application	Notes
Polymer electrolyte or proton exchange membrane	<100°C (<212°F)	50–60%	Polymer membrane (thin plastic film)	100 W to 250 kW per cell	Transportation, stationary	Fast startup, high power density, rapid response to power demand, relatively rugged
Phosphoric acid	160°–220°C (320–430°F)	37–55%; up to 72–80% with heat recovery	Concentrated phosphoric acid	25–250 kW per cell	Stationary	Fuel of choice is natural gas
Solid oxide	800°–1,000°C (1,500–1,800°F)	45–65%; up to 70–85% with heat recovery	Solid nonporous ceramic materials	200 W per cell; 300 kW to 3 MW per module	Stationary, utility	Typically applied in stacks of hundreds; a plant might produce up to 10 MW
Alkaline	23°–250°C (70–482°F)	50–60%	Potassium hydroxide solution (35–50% KOH)	2.2 kW	Spacecraft	Being developed for other applications
Molten carbonate	650°–660°C (1,200°F)	45–60%; 70–85% with heat recovery	Melted carbonate salt mixture	250 kW to >1 MW	Stationary, utility	Corrosive electrolyte limits durability

*Without recovery of cogenerated heat, unless otherwise noted.

Fuel cell applications

Fuel cells produce direct-current (d-c) electricity, but alternating-current (a-c) is the standard for most homes and buildings. In stationary applications, an inverter is needed to change the fuel cell's power from d-c to a-c (the inverter is also called a power conditioner). Most of today's stationary fuel cells use natural gas as their fuel instead of hydrogen, so an onsite reformer is part of the package. The reformer, or fuel processor, extracts hydrogen from natural gas or other hydrocarbon fuels.

When used in a car or other vehicle, the fuel cell supplies power to an electric motor, similar to a battery-powered car. However, a battery merely stores energy instead of generating it, so it must be recharged regularly. In contrast, fuel cells generate electricity from hydrogen fuel, and the vehicles are refueled instead of recharged. Although battery technology is improving, fuel cell vehicles can typically travel farther before refueling than a battery-powered vehicle and can be refueled more quickly than batteries can be recharged.

Some fuel cell vehicles are called hybrids because they incorporate a battery as well as a fuel cell and electric motor. These fuel cell hybrids should not be confused with gasoline-electric hybrid vehicles, which are powered by a conventional gasoline-fueled internal-combustion engine and a battery-powered electric motor.

The fuel cell-motor system in a vehicle includes power-electronics components that integrate the fuel cell with the vehicle's propulsion system, allowing electricity to power the wheels. These electric-drive system components include an inverter and transmission, as well as air, fuel, and control systems.

In most of today's fuel cell vehicles, hydrogen is carried onboard, usually as a compressed gas, so the system doesn't need a reformer to extract hydrogen. If natural gas or other hydrocarbons are the source of the hydrogen fuel, they are usually reformed onsite at the fueling station, since transporting hydrogen is expensive.

Much research was conducted to determine whether reforming hydro-carbon fuels onboard the vehicle might be economical, and a few fuel cell vehicles reform natural gas, methanol, or even gasoline from onboard tanks, rather than carrying hydrogen fuel itself.

How Hydrogen Is Used Today

Hydrogen already has many practical applications in our industries and research laboratories. Worldwide, about 16 trillion cubic feet (tcf) of hydrogen is consumed each year (450 billion cubic meters, or bcm), with the United States accounting for about 20%. Most industrial hydrogen is sold by volume (cubic feet or cubic meters), but hydrogen production can also be measured by weight. Globally about 50 million tons (45 billion kilograms) of hydrogen is produced each year.

Two applications—manufacturing ammonia and refining crude oil—use the greatest amounts of hydrogen by far. Each of these markets in the United States alone consumes more than 1.2 tcf/yr (34 bcm/yr). Ammonia (NH_3) is an important fertilizer and industrial chemical, and hydrogen is a critical ingredient in refinery processes such as hydrocracking and hydrogenation, which produce gasoline, heating oil, and other petroleum products.

In the food industry, edible fats and oils such as margarine and vegetable shortening are solidified using hydrogen. Manufacturing of numerous other commercial products requires hydrogen—dyes, drugs, plastics, vitamins, cosmetics, soaps, lubricants, and cleaners.

Significant quantities of hydrogen are used in manufacturing methanol and electronic components. Hydrogen is also used to recover pure metals from their compounds and to generate very high temperatures for arc welding and glass cutting. In the electric power industry, hydrogen reduces friction in rotating equipment and is used to cool large generators, motors, and frequency changers.

The exploration of outer space by America and other industrial nations depends on substantial amounts of liquid hydrogen as a rocket fuel. Today, the U.S. National Aeronautics and Space Administration (NASA) is one of the largest users of liquid hydrogen in the world.

Nearly all of the hydrogen produced today is consumed onsite at oil refineries and chemical plants, as opposed to being sold on the market. Some applications that require relatively small amounts of very pure hydrogen, such as the electronics industry, use electrolyzers to produce high-purity hydrogen at their facilities.

Hydrogen's Early History

Hydrogen was isolated in the mid-1600s by Robert Boyle, who called the gas factitious air. He produced hydrogen by dropping iron nails into a flask of sulfuric acid (H_2SO_4), then inverting the flask and submerging it in more acid. As the iron and sulfur reacted, hydrogen collected at the top of the inverted flask. Boyle's European contemporaries were also busy exploring the qualities of air and other gases.

More than a century later, Henry Cavendish identified hydrogen as a chemical element, which he called inflammable air in 1766. To produce the gas, Cavendish dropped pieces of zinc, iron, and tin into hydrochloric acid (HCl) and dilute sulfuric acid. He distinguished hydrogen from ordinary air and determined that it is extremely light compared to other gases.

In Cavendish's time, many people still believed that air and water were two of nature's four basic elements, along with fire and earth. But Cavendish found that when hydrogen is combined with air and ignited, the mixture explodes and creates water. He even figured out the approximate proportions of hydrogen and air that are needed to form water, which anticipated knowledge of water's chemical formula, H_2O.

Cavendish's discovery that water is composed of two distinct gases corrected the mistaken idea that water itself was an element. However, water's oxygen component was not recognized for another 20 years, when Antoine-Laurent Lavoisier explained how oxygen combines with burning substances and increases their weight. This French chemist gave hydrogen its modern name, which means water-producer.

Hydrogen's buoyancy led to its use as a substitute for hot air in aeronautical balloons. Jacques A. C. Charles flew a hydrogen-filled balloon to an altitude of almost two miles in 1783. Then in 1794, French army workers built a furnace that used steam and iron to produce hydrogen from water. Gaseous hydrogen was used in 1820 as fuel for one of the earliest internal-combustion engines.

During the 1800s chemists were liquefying many common gases, but hydrogen resisted even the more sophisticated liquefaction efforts of the age until 1898. James Dewar finally succeeded by cooling the hydrogen via expansion (the Joule-Thomson effect), and then adapting a process invented by Karl von Linde in which compressed gas is cooled repeatedly until very low temperatures are reached.

Exploiting Hydrogen's Properties

The idea of an industrialized economy based on hydrogen was conceived by many scientists but expressed perhaps most completely by John B. S. Haldane in the 1920s. He not only envisioned hydrogen as a future fuel, but also brought attention to its energy storage potential, imagined hydrogen's integration with wind energy, and anticipated the costs and social advantages of making a transition to hydrogen.[3]

During the 1920s and 1930s hydrogen began to be exploited for aviation, for example in Germany's Zeppelin dirigibles, which flew travelers on holiday across the Atlantic. Ingeniously, the aircraft's engineers used hydrogen not only to control buoyancy but also as a booster fuel. Within the next two decades, researchers were trying to use hydrogen fuel in cars, trucks, trains, and submarines. The U.S. Air Force had a secret, multimillion dollar program during the 1950s, code-named Suntan, to develop hydrogen as a fuel for airplanes.

In the infamous *Hindenburg* accident of 1937, the German airship caught fire while landing, killing 37 passengers. Most people today associate this incident with hydrogen, which was widely blamed. However, recent research has shown that the fire started on the airship's cloth fabric skin before spreading to burn the onboard diesel fuel and then the hydrogen (see chapter 6).

Until 1932, only one type of hydrogen atom—containing one proton and one electron—was recognized. Then Harold Clayton Urey discovered deuterium, or heavy hydrogen, which also contains a neutron (a subatomic particle that weighs the same as a proton but has no electrical charge).

Hydrogen's third form, the radioactive isotope tritium, was artificially created two years later by adding a second neutron. The extraordinary destructive power of the hydrogen bomb results from the thermonuclear fusion of this hydrogen isotope, not the common form of hydrogen.

Turning hydrogen into a solid metal has been called the Holy Grail of high-pressure physics. Solid crystalline hydrogen was produced in the laboratory in 1979, but solid metallic hydrogen has proved more difficult to create. If it ever becomes practical, solid hydrogen metal holds promise as an ultra-energetic fuel and as a zero-resistance electrical conductor.

An extremely thin layer of frozen hydrogen was compressed into a solid semi-metal in 1989 by Ho-Kwang (David) Mao and Russell J. Hemsley, using enormous pressures more than 2.5 million times greater than the earth's atmospheric pressure. A few years later, scientists at Lawrence Livermore National Laboratory created liquid metallic hydrogen for less than a microsecond, using high temperatures and pressures. But solid metallic hydrogen still remains more of a laboratory curiosity than a practical reality.

Fuel Cells Enter the Picture

Around 1800, the invention of electrolysis—using an electric current to decompose water into its hydrogen and oxygen components—paved the way for the invention of the fuel cell. Many scientists were trying to reverse the process of electrolysis and generate an electric current, but it was William Grove who built a workable fuel cell in 1839 (Fig. 3–3).

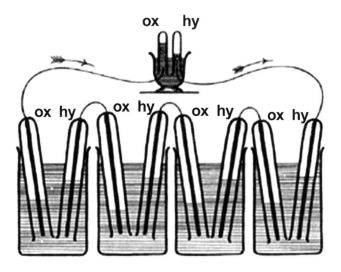

FIG. 3–3. EARLY FUEL CELL. WILLIAM GROVE OPERATED WHAT'S RECOGNIZED AS THE FIRST FUEL CELL IN 1839.

Grove's device combined gaseous hydrogen and oxygen using platinum plates as the electrodes and sulfuric acid as the electrolyte. William W. Jaques, who later substituted phosphoric acid in the electrolyte bath, coined the term fuel cell. Research in Germany during the 1920s laid the groundwork for development of molten carbonate and solid oxide fuel cells.

Grove connected his cells in series to produce more power, as is common today, but what we consider the modern fuel cell was developed by Francis T. Bacon in 1932. Bacon not only replaced the platinum electrodes with nickel, which is much cheaper, but he also compressed the gas. By 1959 his Bacon fuel cell was generating five kilowatts of electricity. Larger versions soon followed, applied in a tractor and a research vessel.

In the 1960s, NASA began using fuel cells to provide onboard electrical power for the Gemini and Apollo spacecraft (Fig. 3–4). The fuel cell's prime advantage in space was its power output per unit weight, about eight times greater than a battery's. NASA's groundbreaking Space Shuttle research in the 1970s and 1980s made possible the resurgence of fuel cell development in the 1990s.

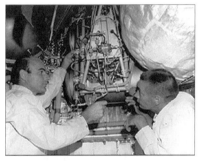

FIG. 3–4. SPACECRAFT FUEL CELL. TECHNICIANS INSPECT A FUEL CELL, WHICH PROVIDED AUXILIARY POWER FOR GEMINI MISSIONS. PHOTO COURTESY OF NASA

Hydrogen's Renaissance(s)

Between World War II and the 1973–74 oil embargo, hydrogen was used widely in oil refineries and as a raw material in manufacturing, as well as in spacecraft. As early as 1970, the term hydrogen economy was coined, reportedly at General Motors. But it was the global energy crisis arising from the oil embargo that spurred scientists to reconsider hydrogen energy and its application in fuel cells.

At the same time, people were more concerned than ever before about the planet's environment, and activists were working globally to reduce the impact of humanity on the earth's ecology. Air pollution from vehicles and power plants became a target for emission control. Chapter 2 discusses these factors and other trends that are once again driving the world toward a hydrogen economy.

The 1970s was an optimistic era for scientists enamored with hydrogen. In fact, one group of enthusiasts came to be called the Hydrogen Romantics. During these years, the International Association for Hydrogen Energy was created, and the world's international hydrogen conference series was initiated.

Federal governments around the world invested in hydrogen research and launched ambitious programs to develop technologies for using hydrogen as well as alternative fuels such as wind power and solar energy. Even nuclear energy was discussed early on as a way to produce hydrogen, until its risks became more apparent.

However, the United States invested less than Europe did, and during the 1980s, as oil prices declined, interest in hydrogen waned. America's low point was 1987, when just $1 million was budgeted for hydrogen R&D. With conservative politicians in charge in America and Britain, the activities of clean air advocates and environmentalists were frustrated.

Hydrogen's second renaissance began in the 1990s, when researchers publicized the global warming trend. This rebirth was heralded by U.S. Vice President Al Gore's book *Earth in the Balance* in 1992 and by Japan's We-Net program in 1993. We-Net grew out of Japan's Project Sunshine, which originated in 1974, and represents the world's first major hydrogen program in response to global climate concerns.[4]

Since then, the European Union has become more aggressive in regulating emissions of carbon dioxide, and countries around the world have agreed to implement global warming control measures. The issue, however, remains controversial (see chapter 2).

An excellent, detailed history of hydrogen appears in *Tomorrow's Energy* by Peter Hoffman.[5] Highlights of international activities during 1987–2000 are summarized in Table 3–2, and recent developments in hydrogen and fuel cell technologies are discussed in this book's chapters 4–7.

TABLE 3–2. RECENT HISTORY OF HYDROGEN AND FUEL CELL R&D*

1987	Government study urges Canada to make hydrogen energy technology a national mission
	1st fuel cell bus project in United States begins with hybrid design
1988	Maiden flight of Soviet Union passenger jet with one engine converted to liquid hydrogen
	First flight of airplane fueled only on liquid hydrogen, a four-seat Grumman flown by William Conrad in Florida
	Soviet Union's high level of hydrogen research activities is revealed at World Hydrogen Energy Conference in Moscow
	Canada launches solid polymer electrolyte fuel cell research program
	Germany begins sea trials of hydrogen fuel cell submarine
	Study begins on shipping electrolytically generated hydrogen from Quebec to Europe

1989	National Hydrogen Association is launched in Washington, DC
	Effort begins to establish international technical standards for hydrogen energy
	BMW builds 1st test stand for hydrogen-powered internal-combustion (i-c) engines
1990	German lab operates world's 1st solar-powered hydrogen production plant
	Soviet Union and West Germany plan to develop propulsion technology for liquid hydrogen jetliner prototype
	Environmental costs of fossil fuel burning are estimated at $2.3 trillion, or $460 per capita worldwide; cost is equivalent to $1/gal gasoline in the US
	Hydrogen exhibits debut at ACHEMA, world's premier triannual chemical equipment industry show in Frankfurt
	GM begins work on 10-kW methanol PEM fuel cell
1991	Tests begin in Germany on liquid hydrogen refueling device for cars and buses
	Fuel cell-battery hybrid, a converted Ford Fiesta mail-delivery car, tested in Pennsylvania
	Mazda unveils HR-X concept car using hydrogen-fueled rotary engine
1992	Grid-independent solar house in Germany uses hydrogen for long-term energy storage
	US Vice President Al Gore releases book, Earth in the Balance: Ecology and the Human Spirit
1993	Ballard rolls out world's 1st PEM fuel cell bus in Canada
	Japan plans 30-year, $2 billion program to promote hydrogen energy internationally via We-Net
	Ballard and Daimler-Benz collaborate on fuel cells for cars and buses
	California funds study of fuel cell locomotives
1994	Funding is won to fuel three pickup trucks in North America on hydrogen produced from solar energy and water
	Daimler-Benz displays 1st NECAR fuel cell vehicle in Germany
	Saudi-German Hysolar demonstration plant begins operation near Riyadh
	Ballard PEM shuttle buses operating in Canada
	Quebec-EU effort unveils 1st of four hydrogen-powered buses
	Europe's 1st fuel cell bus debuts in Brussels, the 87-kW Eureka alkaline cell hybrid
	BMW displays four generations of LH2 fueled i-c engine cars

1995	Chicago tests three hydrogen buses using Ballard PEM fuel cells
	Daimler-Benz plans to convert a commuter airplane engine to liquid hydrogen
1996	Molten carbonate fuel cell plant generates electricity in California using natural gas
	Daimler-Benz NECAR 2 minivan is unveiled in Germany, 1st consumer-friendly, pollution-free fuel cell passenger vehicle
	Toyota debuts experimental SUV using PEM fuel cell
	Liquid hydrogen bus operating in Germany, i-c engine converted from natural gas
	1st hydrogen golf cart delivered in California, using PEM fuel cell
1997	Chrysler shows mockup PEM fuel cell passenger car that would use hydrogen extracted onboard from gasoline
	Hydrogen's blame for Hindenburg disaster is challenged
	Daimler-Benz and Ballard to spend $300 million on developing fuel cells for transportation; Ford joins team
	NECAR 3 is unveiled, to run on 50-kW Ballard PEM fuel cell using methanol
	Daimler-Benz unveils prototype NEBUS using 250-kW PEM fuel cell
	Japan's Musashi Institute of Technology displays LH_2 fueled Nissan station wagon
	Mazda unveils Demio hybrid station wagon using 20-kW PEM fuel cell
	U.S. presidential advisers urge expansion of DOE's hydrogen program
1998	GM and Opel launch global PEM fuel cell project for European cars, Opel shows experimental Zafira minivan using fuel cell
	1st U.S. residential fuel cell is installed, in a ranch home in New York state
	Norwegian conglomerate Norsk Hydro plans electric generation from hydrogen-fueled steam turbines
	Iceland lays groundwork for its hydrogen economy
	Shell sets up renewables division and hydrogen team
	Zevco debuts its alkaline fuel cell Millennium Taxi in London with Shell support
	U.S. DOE publishes 5-year strategic plan for hydrogen program

1999	Shell creates hydrogen division
	Europe's 1st hydrogen gas stations open in Hamburg and Munich
	DaimlerChrysler introduces NECAR 4, shows fuel cell Jeep concept vehicle
	Ford unveils 75-kW fuel cell P2000 research vehicle and mockup of fuel cell SUV, opens hydrogen fueling station at its Michigan research lab
	The Icelandic Hydrogen and Fuel Cell Co. is created by team of Icelandic companies/institutions and three multinationals
	California launches Fuel Cell Partnership with automakers, oil companies, and Ballard
	BMW unveils 5th-generation LH_2 car using liquid hydrogen in i-c engine, and opens refueling station in Munich with a fully automated fuel pump
2000	World's 1st production-ready PEM fuel cell for automotive use is unveiled by Ballard, a 70-kW unit for 1st-generation fuel cell cars
	GM shows fuel cell version of Precept concept car, 105-kW PEM cell
	Ford shows Th!nk FC5 75-kW fuel cell prototype designed for methanol
	Zembus is launched in Belgium, a diesel city bus converted to hydrogen
	Bayernbus I debuts in Munich, two PEM fuel cell buses rolled out
	American researchers describe experimental renewable hydrogen production method based on green algae
	BMW plans world's 1st commercial hydrogen-fueled car, would use internal-combustion engine

*Adapted from Hoffman, pages 46-51, 113-31. Fuel cells and engines noted above operate on hydrogen unless specified otherwise.

References

1. U.S. Department of Energy, Energy Efficiency and Renewable Energy, www.eere.energy.gov/hydrogenandfuelcells/hydrogen/faqs.html.

2. Hoffman, Peter, *Tomorrow's Energy: Hydrogen, Fuel Cells, and the Prospects for a Cleaner Planet*. The MIT Press: Cambridge, MA, 2001. ISBN 0-262-08295-0, page 59.

3. Ibid., pages 30-31.

4. Ibid., page 12.

5. Ibid., pages 19-51.

Applications and Markets

In the future, hydrogen might carry energy to customers like electricity does today. If the world shifts toward hydrogen energy delivery, fuel cells and other applications for hydrogen would serve as the end point, producing power on demand, but with less air pollution than conventional energy systems.

Hydrogen fuel cells are already emerging in stationary applications—plants generating electricity from a fixed location, ranging in power output from a few kilowatts up to the multi-megawatt range. Some of these technologies are fairly mature, but costs remain prohibitively high for widespread use, and fuel cell lifetimes need to be extended.

In transportation markets, the challenge is even greater—mass-produce hydrogen-powered cars and trucks and buses that perform just as well as the most recent showroom models, but which don't sacrifice valuable space for people and their cargo. And the vehicles must be just as quick and easy and safe to fill up and drive.

Hydrogen fuel cells could power these vehicles of the future. The automotive market is a longer shot—but without it, hydrogen might be doomed to go down in history as another failed alternative transportation fuel.

In this chapter, we'll explore applications for hydrogen power and identify some companies that are developing and demonstrating hydrogen technologies, often with government funding. We'll start with the market for hydrogen-powered vehicles, then take a look at stationary micropower plants, and follow up with the application that's closest to mass introduction—laptops, cellular phones, and other portable devices powered by miniature fuel cells.

As you'll see, many of these market players are developing both stationary and automotive fuel cell technologies. Already these two markets are beginning to merge and might eventually become indistinguishable, if fuel cells designed for vehicles can be plugged into homes and businesses to supply peak power, or even into industrial facilities for round-the-clock operation.

Fuel cells produce electricity from hydrogen and oxygen in a clean, flameless process that is roughly 40% to 65% efficient. When heat is recovered and used, fuel cell systems can approach efficiencies of 85%. Fuel cell technologies are categorized by their type of electrolyte, with PEM cells most common in vehicles and phosphoric acid systems in stationary micropower applications (see chapter 3).

A Rapidly Growing Market

The market for equipment related to hydrogen applications is growing rapidly, according to a report by Business Communications Co. Worldwide, the market for hydrogen generation, storage, and delivery devices could nearly double by 2008, increasing at a rate of 15.8%/yr from an estimated $702 million today to nearly $1.5 billion—a tempting target indeed for global entrepreneurs.[1]

The report says that R&D funding will sustain the hydrogen industry for the time being, along with early market entry of small, portable fuel cell-powered devices. After 2010, automotive and stationary hydrogen equipment should dominate the market.

The value of hydrogen production, storage, and dispensing devices for fuel cell purposes was about $321 million in 2002 and is anticipated to increase due to research spending and sales of early commercial units. Costs remain high due to a lack of mass production, which should move into a higher gear soon.

Stationary hydrogen micropower, which accounted for about 66% of the total hydrogen equipment market, should move into full commercial production by 2010, said Business Communications. Automotive systems, which accounted for 28% of the hydrogen equipment market in 2003, could increase its share to 35% by 2008. Only a few companies offer fuel cells as replacements for batteries in portable applications, which were valued at about $39 million in 2003 and should grow to $81.3 million in 2008.

Small, onsite hydrogen generators will emerge as a winner in the market for hydrogen on demand, said a second report by Business Communications. On-demand plants typically supply ultra-pure, electrolytically produced hydrogen for fuel cells, laboratories, and scientific instruments. Revenues from these plants will grow on average by 15.1%/yr during 2003–2008, reaching $83.3 million.[2]

The market for micropower equipment such as fuel cells, micro-turbines, and engine generators rebounded in 2003, said a report by Primen for the Electric Power Research Institute. More than 12,000 businesses in North America were exploring micropower options in the 100-kilowatt to 10-megawatt range and expressed a 50% or greater likelihood of adopting a micropower solution during 2004–2005.[3]

Surprisingly, these strong prospects have realistic expectations for an economic return on micropower projects, with 86% saying they're willing to accept a payback of four years or longer, said Primen. However, in addition to economic savings, micropower equipment suppliers must address user concerns such as equipment maintenance, environmental permitting, and escalation in the price of natural gas fuel. Major blackouts create opportunities to attract micropower customers, but only for a short time because the effects attenuate quickly.

Oil Companies Enter the Hydrogen Market

A few oil/energy conglomerates have become pioneers in promoting hydrogen—such as BP and Shell, both based in Europe, and America's ChevronTexaco. Other U.S.-based multinational oil companies are generally less receptive to hydrogen and fuel cells, although some are partners in a California state hydrogen program (see *The California Hydrogen Highway* in this chapter).

BP involved in automotive hydrogen

This energy multinational changed the meaning of its corporate name from the original British Petroleum to the slogan "Beyond Petroleum." As BP's Lord Browne has said, "We simply cannot survive for long if we remain so out of tune with our consumers' perceptions, and the next generation's attitudes."[4]

BP already produces about 5,000 metric tons of hydrogen per day for industrial uses and is involved in fuel cell vehicle demonstrations in Europe, Asia, Australia, and North America.[5] For example, as a partner with the Chinese Ministry of Science and Technology, BP will build and supply fueling stations for hydrogen vehicle fleets in Beijing and Shanghai, planned to start up in 2005. The Ministry is cofunding the project with the United Nations Development Programme. Additional BP hydrogen projects are described later in this chapter.

BP also is supporting research at Tsinghua University on the role of hydrogen in China, work at Shenyang on storage of hydrogen in carbon materials, and development with the Dalian Institute on direct conversion of methane to hydrogen.

If solar energy as a source of hydrogen for fuel cells becomes economically feasible, BP is well-positioned in the hydrogen production market with its BP Solar International division, a leading manufacturer of solar equipment.

Shell's hydrogen program

As another world oil giant supporting hydrogen, Shell planned to spend up to $1 billion during 2001–2006 on hydrogen energy R&D. Shell's hydrogen division, formed in 1999, is demonstrating hydrogen fueling applications in the three major hydrogen markets—Japan, Europe, and North America.

For example, Shell is GM's partner in providing hydrogen for a fleet of six fuel cell-powered minivans in Washington, DC. The U.S. Postal Service is leasing one of the GM minivans for mail delivery in the area. To fuel the small fleet, Shell and GM opened America's first hydrogen pump at a retail gas station in 2004. The facility, integrated into an existing Shell service station, provides both gaseous and liquid hydrogen.

Shell is also involved in creating Iceland's hydrogen economy (see *Iceland's Ectos project* in this chapter) and is a partner in HERA Hydrogen Storage Systems, a joint venture working on a metal-hydride storage system that will operate at low temperatures compatible with PEM fuel cells.

Shell also supports hydrogen and fuel cell technology through two venture capital entities—Chrysalix Energy Ltd. Partnership, which promotes early-stage companies with high growth potential, and Conduit Ventures Ltd., which focuses on companies with established technologies that are moving toward commercial sales. Shell also manufactures and sells solar energy products such as photovoltaic components and systems.

Shell plans to test a small hydrogen generator designed for home fueling of vehicles with partner Stuart Energy Systems Corp. (formerly Vandenborre Energy Systems). The device is a miniature electrolyzer that uses electricity to convert water to a slow stream of hydrogen. The idea is for people to park their cars when they come home and connect them to the unit for an overnight refill (see *Hydrogen Fueling At Home* in this chapter).

ChevronTexaco teams with Hyundai, UTC

This American energy company will design and build up to six hydrogen fueling stations, including one based on onsite steam reforming of natural gas that will dispense up to 150 kg/day of hydrogen and use any excess to generate electricity from a stationary fuel cell. This hydrogen station, scheduled for operation in 2005, will initially support three fuel cell transit buses in California.

The fuel cell vehicles and fueling stations are part of a five-year demonstration funded by the U.S. DOE. Partner Hyundai Motor Co. will supply a fleet of up to 32 Tucson hydrogen SUVs using fuel cells made by a United Technologies Corp. (UTC) subsidiary. An additional station in the northeastern United States might be built to test performance in a cold climate.

Also, a ChevronTexaco unit was developing its Halias™ technology, which makes hydrogen from natural gas or propane. A Halias fuel processor was selected for a Canadian demonstration project.

Automotive Fuel Cell Market

Of the three primary markets for fuel cells, transportation is the farthest from mass production and widespread use, but hydrogen fuel cell vehicles seem to generate the most excitement, by virtue of their visibility and lavish publicity. Several high-profile demonstration projects are highlighted here and summarized in Tables 4–1 and 4–2 at the end of this section. Appendixes A–C provide additional details on vehicles and fueling stations worldwide.

Most of the fueling stations being installed for these vehicle demonstrations use small, onsite reformers or electrolytic generators to provide hydrogen, so fuel cell vehicles no longer need to carry their own fuel processor onboard to reform gasoline, methanol, or natural gas. Automakers have shifted toward onboard hydrogen storage instead. This has led the industry to wind down its efforts to develop onboard processors for fuel cell vehicles. For example, a joint project called HydrogenSource LLC was dissolved in 2004.

At the same time, competition has heated up from gasoline-electric hybrid vehicles, forcing fuel cell developers to redouble their efforts to extend service life, reduce costs, and improve efficiency. Year-on-year sales of Toyota's gasoline-electric Prius® increased by more than 119% during 2003–04, leaving the company with a backlog of 22,000 orders and allowing it to boost its price for the 2005 model. Monthly production was increased by 50% to meet unprecedented demand, and Toyota planned to double Prius hybrid deliveries to the United States.

Toyota projected hybrid sales of 300,000 vehicles in 2005, including 24,000 of its new Lexus hybrid sport utility vehicle.[6] The Lexus SUV created strong demand in advance of its launch, with more than 9,000 pre-sold orders.

Honda's hybrid Civic sales rose by almost 25% in 2003–04, and Honda introduced a six-cylinder Accord hybrid that achieves the same fuel economy as the four-cylinder Civic but delivers more horsepower and torque.

A hybrid gasoline-electric GMC Sierra pickup truck, introduced in 2004, features 120-volt electrical outlets in the bed and cab, allowing drivers to park the truck and operate power tools, campsite gear, or home appliances. GM calls the truck a 295-horsepower generator on wheels. Ford launched production of a hybrid version of its Escape SUV in 2004 and plans to make 20,000 of the vehicles during the first full production year.

Fuel Cell Vehicle Demonstrations

In 2004 General Motors set a world distance record for fuel cell vehicles, traveling 6,059 miles (9,751 kilometers) and crossing 14 countries, from northern Norway to the Portuguese capital of Lisbon, the continent's most westerly point. GM's goal is to make fuel cell vehicle technology commercially viable by 2010.

The record was set by GM's HydroGen3 fuel cell prototype (Fig. 4–1), which is based on the Opel Zafira minivan. Previously, the record was held by a DaimlerChrysler fuel cell car, which traveled 3,281 miles (5,280 km) from San Francisco to Washington DC in 2002.

FIG. 4–1. GM HYDROGEN3 IN MARATHON. THIS FUEL CELL VEHICLE PROTOTYPE, BASED ON THE OPEL ZAFIRA MINIVAN, TRAVELED MORE THAN 6,000 MILES THROUGH EUROPE (9,700 KM), SETTING A WORLD DISTANCE RECORD. PHOTO COURTESY OF GM.

The GM fuel cell prototype minivan withstood extreme temperatures and road conditions during its 38-day journey and performed without any unscheduled stops for repair. The vehicle was subjected to high speeds, steep mountain passes, and temperature differences of more than 32°F (18°C).

A hydrogen fuel cell powered the minivan's 60-kW electric motor. The front-wheel drive vehicle accelerated to 100 mph (160 km/h) and averaged about 300 miles per day (480 kilometers). A mobile fueling station was used to replenish the hydrogen during the trip. On average, refueling took about four minutes.

GM is also developing the Sequel concept car (Fig. 4–2), which can accelerate to 60 mph in less than 10 seconds and achieves a 300-mile traveling range (close to 500 km).[7]

FIG. 4–2. GM SEQUEL CONCEPT CAR. TRAVELING RANGE AND ACCELERATION WERE GREATLY IMPROVED IN THIS HYDROGEN FUEL CELL CONCEPT CAR. PHOTO COURTESY OF GM

The Sequel carries three fuel tanks containing gaseous hydrogen at 10,000 psi, and the fuel cell stack provides 73 kW of power. The Sequel's propulsion and control systems are packaged in an 11-inch-thick chassis. Traditional mechanical systems are replaced with electronically controlled drive-by-wire technology.

Ford Focus fleet

Ford Motor Co. began production of a fleet of fuel cell-powered Focus sedans in late 2004, which will be deployed in several cities worldwide. The Focus is one of six prototype fuel cell vehicles in development. One vehicle has been in real-world, limited use, serving as a delivery truck for the U.S. Postal Service at various locations.[8]

Ford and partner BP will demonstrate the fleet of up to 30 hydrogen fuel cell vehicles in three U.S. cities—Sacramento, CA; Orlando, FL; and the Detroit, MI area—as well as in Vancouver, British Columbia, and Berlin. Ten Focus fuel cell sedans (Fig. 4–3) will operate in the Sacramento area beginning in 2005, fueling up at stations to be built and operated by BP, as well as an existing station at the University of California-Davis campus nearby.[9]

Fig. 4–3. Ford Focus fuel cell sedan. Up to 30 of these hydrogen-powered cars are being tested in Sacramento, CA; Orlando, FL; and Detroit, MI beginning in 2005. Photo courtesy of Ford.

The Sacramento-area vehicles will be operated by California state and local agencies, including the Air Resources Board, California Energy Commission, the Department of General Services, the City of Davis, and by the local utility, Sacramento Municipal Utility District. The Ford-BP project is funded by the U.S. DOE as part of its FreedomCAR program.

The four-door Focus fuel cell sedan can travel about 100 miles (160 km) between fill-ups at maximum speeds of about 80 mph (130 km/h). The electric motor features a peak efficiency of 91% and provides 67 kW of power from its PEM fuel cell, made by Ballard Power Systems. A 3,600-psi tank supplies compressed hydrogen fuel to the cell.

Ford is also the largest shareholder in Mazda, which sells rotary engine-powered cars. After developing prototype hydrogen-fueled rotary engines in the past, including a hydrogen-powered Miata sports car, Mazda was planning to continue this work and also collaborate with Ford on fuel cell technology.

Honda FCX cars

Two Honda FCX hydrogen fuel cell-powered cars were delivered in 2004 to the City of San Francisco, which planned to build refueling infrastructure to support their daily operation. The four-passenger vehicles (Fig. 4–4) have a range of up to 160–170 miles (260–270 km).

FIG. 4–4. HONDA FUEL CELL CAR. TWO FCX HYDROGEN-POWERED CARS ARE OPERATING IN SAN FRANCISCO, AND HONDA IS TESTING THE COLD-WEATHER PERFORMANCE OF THE FCX VEHICLES' FUEL CELL STACK IN NEW YORK STATE AND IN JAPAN. PHOTO COURTESY OF HONDA.

Honda was also leasing two FCX fuel cell vehicles to California's South Coast Air Quality Management District. Altogether, 12 Honda fuel cell cars were on the road in America in 2004, including five in Los Angeles. The FCX meets U.S. federal safety standards and is certified for everyday commercial use by the U.S. Environmental Protection Agency (EPA) and California Air Resources Board.

The FCX, with a maximum speed of 93 mph (150 km/h), is powered by a 78-kW Ballard PEM hydrogen fuel cell. The electric motor delivers 60 kW of power. The 5,000-psi storage tank (350-bar) holds 156.6 liters of compressed hydrogen gas, for a range of more than 170 miles (270 km). Electricity is stored in a Honda ultracapacitor, an electrostatic device that is beginning to replace batteries in some fuel cell vehicles.

Honda has developed a fuel cell stack that can start up in sub-zero temperatures (as low as −4°F, or −20°C), a breakthrough on the path toward mass markets. The stack's performance was successfully demonstrated in 2004. The fuel cell stack will be tested on the road in Honda's FCX vehicle in New York state and in Japan.

Honda's 2005 fuel cell FCX produces 33% more power than the 2004 model—107 horsepower versus 80—and improves fuel efficiency by 20%, achieving the equivalent of 57 mpg. This translates to a travel range of 190 miles (305 km). The new FCX model is also equipped with Honda's low-temperature fuel cell stack.

DaimlerChrysler cars, trucks, and buses

During 2004 United Parcel Service (UPS) was operating a DaimlerChrysler F-Cell car on a daily express-delivery route around Ann Arbor, MI. The vehicle is based on a Mercedes-Benz A-Class. UPS fuels the vehicle at the EPA's National Vehicle and Fuel Emissions Laboratory.

Positive road-test results led to plans for three hydrogen fuel-cell versions of the Dodge Sprinter delivery truck—America's first medium-duty fuel cell delivery vehicles (Fig. 4–5). UPS began operating the first truck in Los Angeles in 2004. The new Sprinter trucks feature a 20% increase in powertrain efficiency, a 40% increase in range to 155 miles (249 km), and a 45% increase in peak engine power.

FIG. 4–5. HYDROGEN-POWERED DELIVERY VEHICLES. FOLLOWING TESTS ON A FUEL CELL DELIVERY CAR, UPS BEGAN OPERATING AMERICA'S FIRST MEDIUM-DUTY FUEL CELL DELIVERY TRUCK IN 2004. TWO MORE HYDROGEN-FUELED TRUCKS WERE TO ENTER SERVICE IN 2005. PHOTO COURTESY OF UPS.

DaimlerChrysler also began delivering fuel cell cars to German fleet customers in 2004. In these four-passenger A-Class vehicles (Fig. 4–6), the fuel cell system and hydrogen cylinders are located under the floor, allowing more space for people and cargo. A DaimlerChrysler-Ballard fuel cell system was also used by Mitsubishi Motors Corp. in its Grandis minivan, which was being road-tested in Japan.

Fig. 4–6. DaimlerChrysler fuel cell cars. These F-Cell A-Class cars were delivered to fleet customers in Berlin in 2004. Photo courtesy of DaimlerChrysler.

Similar A-Class fuel cell vehicles are also being delivered to southern California for operation by the South Coast Air Quality Management District. The cars have a range of about 100 miles (160 km) and top speed of 85 mph. Power from the 65-kW electric motor is comparable to 88 horsepower, enabling acceleration from 0 to 60 mph in 16 seconds.

DaimlerChrysler introduced its first fuel cell vehicle in 1994 and is providing 30 Citaro fuel cell buses for demonstrations throughout Europe (see *Europe's* CUTE *Bus Program* in this chapter). In 2005, a total of 60 DaimlerChrysler fuel cell vehicles will be operating in Japan, Germany, Singapore, and the United States, including 20 to be tested in Michigan and California.

With partner BP, DaimlerChrysler is participating in a demonstration project of six fuel cell cars in Singapore. BP added hydrogen fueling facilities at one of its existing stations there in 2004 and was considering opening a second. The two-year project is supported by Singapore's Economic Development Board.

Three more DaimlerChrysler Citaro fuel cell buses will be delivered to Beijing under an agreement with China's Ministry of Science and Technology. The buses will operate from 2005 through 2007. BP will build stations and supply hydrogen to fuel cell vehicles in Beijing and Shanghai, as part of China's demonstration projects at the 2008 Olympics and the 2010 World Expo.

Ballard's automotive fuel cell research

A strategic alliance between Ballard Power Systems and automakers DaimlerChrysler and Ford has resulted in PEM fuel cells being used in Citaro buses, Mercedes and Focus cars, and other fuel cell vehicles being demonstrated. Ballard's PEM products are powering the largest number of fuel cell vehicles in the world—more than 110, including 45 buses.

Over the past decade Ballard's most important contributions to hydrogen fuel cell technology, says Vijay Vaitheeswaran in *Power to the People*, have been shrinking the size of the stack required to power a car—"from the size of a refrigerator to the size of a microwave"—as well as dramatically reducing the amount of expensive platinum required and redesigning the fuel cell stacks to be easier to manufacture and assemble.[10]

Ballard sold its German vehicle fuel cell business in 2004 to DaimlerChrysler and Ford, which will develop and manufacture system components except for the fuel cell itself, while Ballard will focus on next-generation automotive fuel cells and electric drives. Ballard also develops portable and stationary fuel cell appliances and micropower plants (see *Residential and commercial micropower* in this chapter).

Toyota SUV, Prius

Toyota has developed a hydrogen fuel cell-battery hybrid vehicle based on its Highlander mid-sized sport utility vehicle. The SUV uses a proprietary fuel cell system that generates a peak of 90 kW, which powers a 109-hp electric motor (194 lb-ft of torque). Hydrogen is stored onboard in four 5,000-psi tanks (350-bar). Nickel-metal hydride batteries store the fuel cell's power and feed it on demand to the motor drive.

A fuel cell Highlander was delivered in 2004 for testing at the University of California-Irvine's National Fuel Cell Research Center, where a Stuart Energy Station will provide the hydrogen fuel via water electrolysis. The five-passenger SUV has a range of up to 180 miles (290 km) and a top speed of 96 mph (154 km/h).

Toyota is also working with Energy Conversion Devices, Inc. (ECD Ovonics) on a hybrid fuel cell Prius car using Texaco Ovonic Hydrogen Systems' metal hydride technology, which stores hydrogen onboard in a solid form. The technology is being developed by Texaco Ovonic for automotive applications and for stationary storage of bulk hydrogen, as well as portable storage for plugging into a growing array of hydrogen-fueled devices (see *Uninterruptible power supply applications* in this chapter). Partner ChevronTexaco owns a 50% stake in Texaco Ovonic.

A metal hydride is formed when gaseous hydrogen molecules dissociate into individual hydrogen atoms and bond with metal atoms in the storage alloy. Removing heat drives this absorption process, while adding heat reverses the chemical reaction, causing the hydrogen atoms to reform as H_2 molecules inside the storage vessel. Metal hydride storage density is higher than that of compressed gaseous or liquid hydrogen (see chapter 6).

The fuel cell Prius with metal hydride storage can travel nearly 150 miles (240 km) and takes about 10 minutes to refuel at a pressure of 1,500 psi (105 bar). The 50-liter storage vessel contains 3 kilograms of hydrogen (equivalent to about 3 gallons of gasoline) at an operating pressure typically less than 200 psi (14 bar). The vehicle's range could be doubled by using two storage vessels, and a more efficient heat exchanger could shorten refueling time, said Ovonic.

Texaco Ovonic's metal hydride storage will also be used in six hybrid hydrogen-electric sedans in 2005. The converted vehicles will be operated by California's South Coast Air Quality Management District. The first modified sedan began test drives there in 2004.

Other fuel cell vehicles

Nissan's fuel cell SUV. A hybrid fuel cell vehicle based on Nissan Motor Co.'s X-Trail SUV was leased by Japan's Cosmo Oil Co., which also operates Toyota and Honda fuel cell vehicles. Nissan developed the fuel cell-battery X-Trail with improvements over earlier models, including maximum power of 85 kW and a driving range of 220 miles (350 km). The vehicle, which uses a lithium-ion battery, is approved for public use on Japanese roads.

Nissan plans to develop its own automotive fuel cell and unveil a new vehicle in 2007 for leasing to Japanese and U.S. government agencies. Meanwhile, an aftermarket fuel cell-battery hybrid version of a Nissan Frontier pickup truck was offered commercially in the United States by Anuvu, Inc.

European automaker to test novel storage system. PSA Peugeot Citroën plans to explore fuel cell vehicles using the Hydrogen on Demand® fuel system developed by Millennium Cell Inc. Peugeot was testing two of the systems in 2004 with support from French utility Gaz de France.

The fuel storage system generates hydrogen from sodium borohydride, which is derived from sodium borate, commonly known as Borax. Dissolved in water and passed through a proprietary catalyst chamber, the sodium borohydride releases a stream of pure hydrogen. The by-product is water and Borax (see chapter 6).

Fuel cell scooter. Honda has developed a motor-scooter powered by a fuel cell located in the center of the vehicle (Fig. 4–7). With further design refinements, the scooter could achieve the same range, as well as comparable storage space, as the 125-cc engine-powered version popular with commuters worldwide. Honda miniaturized its fuel cell stack for this application.

Fig. 4–7. Honda scooter. This fuel cell-powered motor-scooter could achieve the same range, as well as comparable storage space, as the 125-cc engine-powered version popular with commuters worldwide. Photo courtesy of Honda.

Alkaline fuel cell golf cart. Alkaline technology, though not as widespread, is entering use in a hydrogen-powered golf cart that can cruise at up to 19 mph (31 km/h). The prototype, made by Astris Energi Inc., uses a 1-kW alkaline fuel cell and a 33-liter, refillable carbon-fiber tank for hydrogen storage. Acceleration and hill-climbing is superior to conventional battery models, said Astris, and the cart can run for up to three days under normal usage.

Astris also offers a 2.4-kW alkaline fuel cell for portable power applications. Electrical efficiency can exceed 50%.

Tractor manufacturer testing fuel cell hybrid. John Deere's ePower Technologies was designing and building technology demonstration vehicles to determine whether they have advantages for its customers. The ePower team has created a hydrogen fuel cell-based work vehicle designed for outdoor or indoor use that can operate for six hours before refueling.

The vehicle uses a 20-kW PEM fuel cell (the Hydrogenics HyPM-LP[2] power module), a nickel metal-hydride battery, compressed hydrogen storage at 5,000 psi (350 bar), and twin 30-kW electric motors. However, Deere has no commercialization plans for the technology.

UTC partnerships. Nissan and South Korea's Hyundai Motor Co. could use PEM fuel cell technology being developed by UTC Fuel Cells. Hyundai expects to place fuel cell vehicles into fleet applications by 2005, with limited consumer availability planned for 2010. UTC also has agreements to work on automotive fuel cells with BMW AG and Renault.

BMW sticks with engines

Unlike most of its fellow automakers, BMW plans to offer a hybrid gasoline-hydrogen car based on an internal-combustion engine rather than a fuel cell. Although BMW has explored PEM fuel cells with UTC and has tested fuel cell vehicles, it believes that they can't deliver a high-performance driving experience. Several other automakers have also worked on hydrogen-fueled engines, including Nissan, Ford, and DaimlerChrysler.

BMW's plans are to adapt its 7-Series cars to hybrid operation. The company is also developing a new breed of race cars adapted to run on liquid hydrogen fuel. The prototype H2R has set nine international records, including sustained speed of more than 180 mph over a mile-long stretch. The H2R features a 6.0-liter, V-12 engine that generates 232 horsepower.

BMW has tested hydrogen-fueled internal-combustion engines for many years using liquid hydrogen, or LH_2, and built the first test stand for them in 1989. BMW also participated in developing the hydrogen fueling station at the Munich airport, the world's first public station to offer liquid hydrogen. The fully automated pump allows drivers to swipe their card and activate a robotic arm that swings out and locks into the car's fueling port. The device can fill the tank in less than two minutes (see chapter 7).

Europe's CUTE Bus Program

One of Europe's most prominent hydrogen programs, CUTE, is demonstrating 30 fuel cell-powered public buses, three in each of ten cities— Amsterdam, Barcelona, Hamburg, London, Luxembourg, Madrid (Fig. 4–8), Porto (Portugal), Reykjavik (Iceland), Stockholm, and Stuttgart (Germany).

FIG. 4–8. CUTE HYDROGEN FUEL CELL BUS. THIS DAIMLERCHRYSLER FUEL CELL BUS IN MADRID IS ONE OF 30 BEING TESTED THROUGHOUT EUROPE AS PART OF THE CUTE DEMONSTRATION PROGRAM. PHOTO COURTESY OF EU CUTE.

The Citaro urban transit buses, which generate 280 kW from compressed hydrogen stored on the roof, are made by a DaimlerChrysler unit. Buses and other fleet vehicles make good candidates for hydrogen fuel cell demonstrations because they are usually parked overnight at a central location with its own fueling facilities.

The European Union calls this the first project worldwide to address simultaneously hydrogen production, fueling in city centers, and operation in commercial public transport systems. The buses are being driven just like conventional ones, on the same routes and schedules, and the entire hydrogen system will be tested, from production to passenger reaction and acceptance.[11]

The CUTE bus demonstration project, which was launched in 2001, runs through 2006. Performance under varying weather conditions and topography are being monitored, and environmental benefits will be quantified. Shell is a partner in the Amsterdam, Luxembourg, and Reykjavik bus projects. In Amsterdam, where the station opened in 2003, the challenge is dense traffic, and in Luxembourg the steep, hilly terrain.[12]

London received its first bus in 2003, and the Portugal fueling station opened in 2004. BP is a partner in the London venture and also provided fueling stations for CUTE's bus demonstrations in Barcelona, Hamburg, and Stuttgart.

At the Hamburg fueling station, Norwegian energy company Norsk Hydro provided the electrolyzer that produces hydrogen fuel from water. The unit was installed by Germany's HEW (Hamburgische Electricitäts-Werke), which gained experience with electrolyzers during the Munich airport hydrogen project.

Japan's Hydrogen Fuel Cell Program

The Japan Hydrogen Fuel Cell (JHFC) demonstration project is Japan's first extensive research that involves actual operation of fuel cell vehicles for demonstration purposes. In 2003 fuel cell vehicles from eight car manufacturers and fuel cell buses for commercial routes participated in trial runs on highways. Highway performance and station usage data were obtained for evaluation, including drivability, reliability, environmental characteristics, and fuel consumption.[13]

As part of the JHFC project, three of GM's HydroGen3 fuel cell cars were tested in the Tokyo area by Federal Express. The vehicles can travel 250 miles (400 km) at speeds up to 100 mph (160 km/h).

The JHFC project is directed by Japan's METI, which not only supports R&D but also seeks to raise public awareness. Japan's project includes operation of five hydrogen fueling stations around Tokyo, one of them to be built and operated by a Shell unit.

The JHFC project is also called the world's first parallel operation of hydrogen stations based on several different fuel production and storage technologies: reforming of desulfurized gasoline, naphtha, liquefied petroleum gas (LPG), methanol, petroleum, and city gas, as well as storage of compressed and liquid hydrogen and electrolysis of lye (an industrial chemical). Japan had 10 fueling stations operating in 2004.

Iceland's Ectos Project

A joint venture called Iceland New Energy Ltd. is developing hydrogen production technologies and fuel cell applications for Iceland, which plans to be the world's first hydrogen-powered economy before 2050. Shell, DaimlerChrysler, and Norsk Hydro are partners with Iceland in the ambitious undertaking.

To explore hydrogen's potential, the venture set up the Ectos project (Ecological City Transport System), and a Shell hydrogen fueling station was installed in Iceland's capital of Reykjavik (Fig. 4–9). Norsk Hydro supplied an electrolyzer, and DaimlerChrysler manufactured the island's three hydrogen-powered city buses, which comprise 4% of Reykjavik's bus fleet. The buses began operating in 2003 and are being tested for two years in cooperation with Europe's CUTE project.

FIG. 4–9. ICELAND'S HYDROGEN FUELING STATION. ONE OF THREE FUEL CELL BUSES FILLING UP AT SHELL'S HYDROGEN STATION IN REYKJAVIK. PHOTO COURTESY OF SHELL.

California Fuel Cell Vehicle Programs

California leads the U.S. states in hydrogen fuel cell vehicle and infrastructure initiatives. Many of the state's demonstration projects center around Los Angeles, its largest and most polluted city, and Sacramento, its capital.

The California Hydrogen Highway

California's Hydrogen Highway program, initiated in 2004, intends to catalyze the state's transition to a clean, hydrogen-based transportation economy by building an early network of 150 to 200 hydrogen-fueling stations (approximately one station for every 20 miles of major highway), in addition to California's existing 11 stations (Fig. 4–10). The initial, low-volume fueling network is estimated to cost $90 million, mainly from private investment, and should be achievable by 2010.[14]

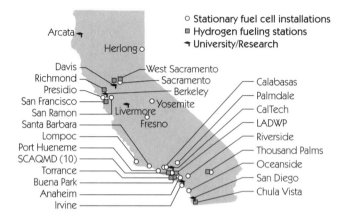

FIG. 4–10. CALIFORNIA'S HYDROGEN AND FUEL CELL FACILITIES. CALIFORNIA
PLANS TO ADD UP TO 200 MORE HYDROGEN VEHICLE FUELING STATIONS TO ITS
EXISTING 11 FACILITIES. THE STATE ALSO HAS MORE THAN 20 STATIONARY FUEL
CELL INSTALLATIONS. ARTWORK COURTESY OF THE STATE OF CALIFORNIA.

Partners in the California Hydrogen Highway, which total about 90 companies, public agencies, and research organizations, include several major
automakers, engine manufacturers, and hydrogen fuel cell system suppliers,
as well as BP, Shell, ChevronTexaco, and ExxonMobil.

As part of the hydrogen highway program, California opened the first
new fueling station in 2004 at the South Coast Air Quality Management
District's headquarters in Diamond Bar. The next three stations are planned
for Burbank, Riverside, and Santa Ana.

California Fuel Cell Partnership

The California Fuel Cell Partnership (CaFCP) is a collaboration of 30
members including auto manufacturers, oil and energy companies, fuel cell
technology developers, and government agencies. Since it was formed in
1999, the group has fielded more than 55 fuel cell vehicles that have logged
more than 145,000 miles (233,000 km).

CaFCP's 2004 Road Rally featured fuel cell vehicles from DaimlerChrysler, Ford, General Motors, Honda, Hyundai, Nissan, Toyota, and Volkswagen. Through 2007, the CaFCP will facilitate placement of up to 300 more fuel cell vehicles in fleet demonstration projects, including buses in three transit districts.

SunLine bus demonstration

Southern California's Coachella Valley is home to what's been called the world's most complex hydrogen demonstration project. California's first fuel cell-battery hybrid bus entered service there in 2002, a 30-foot Thor Thunder-Power model powered by a 75-kW UTC fuel cell and operated by SunLine Transit Agency, which serves an area of 1,100 square miles (2,800 square kilometers), including nine participating cities and Riverside County.

Buses running on hydrogen and hydrogen-natural gas mixtures are used for public transport and fueled at SunLine's public-access facility. Two different types of electrolyzers, one supplied by a photovoltaic grid, produce hydrogen onsite. The facility attracts more than 5,000 visitors a year from around the world.

Previously, SunLine established itself as an alternative vehicle fuel test bed by switching its fleet to compressed natural gas, then moved toward the hydrogen-natural gas blends and fuel cell buses. Unattended, 24-hour operation of a hydrogen generator at SunLine's fueling station was approved in 2004 by the local fire department.

U.S. FreedomCAR Program

In DOE's FreedomCAR program, a partnership of DaimlerChrysler, Ford, and GM represents the automotive industry. FreedomCAR envisions a hydrogen-powered U.S. transportation system and conducts research on lightweight materials, clean combustion, and power electronics. Technical teams are investigating fuel cell technologies, advanced combustion and emissions control, hydrogen storage and vehicle interface, and other research areas.

Hydrogen and **Fuel Cells:** A Comprehensive Guide

TABLE 4–1. DEMONSTRATION AND DEVELOPMENT OF HYDROGEN VEHICLES

Automaker	Vehicles	Partners	Locations	Fueling Stations	Dates	Notes
BMW	7-Series car					Internal-combustion engine runs on liquid hydrogen
Daimler-Chrysler	Thirty 280-kW Citaro urban transit buses	European Union CUTE project	Amsterdam; Barcelona; Hamburg; London; Luxembourg; Madrid; Porto, Portugal; Reykjavik; Iceland; Stockholm; Stuttgart, Germany	BP, Shell	2001–2006	
	Mercedes-Benz A-Class fuel cell delivery car	United Parcel Service	Ann Arbor, MI	U.S. EPA facility		
	Three Dodge Sprinter fuel cell delivery trucks	United Parcel Service	Los Angeles, Sacramento, Ann Arbor, MI	Existing	2004–2005	1st U.S. medium-duty package delivery trucks using hydrogen fuel cells
	Mercedes-Benz fuel cell A-Class	German fleet customers	Berlin	Public station available	2004	
	Three 205-kW Citaro buses	China's Ministry of Science and Technology	Beijing	BP	2005–2007	
	Six cars	Singapore's Economic Development Board	Singapore	BP		
Ford	Up to 30 67-kW Focus passenger cars	BP, U.S. DOE/FreedomCAR	Sacramento, CA (10); Orlando, FL, and Detroit, MI	To be built by BP; one existing at UC-Davis	2005	Sacramento vehicles operated by state/city government and local utility
	Fuel cell delivery truck	U.S. Postal Service	Various U.S.			Real-world, limited use.

Company	Vehicle	Partner	Location	Station	Year	Notes
GM	Six HydroGen3 fuel cell minivans	Shell, U.S. Postal Service	Washington, D.C.		2004	U.S.'s 1st hydrogen pump at retail station
	HydroGen3 fuel cell minivan	Linde AG	Europe marathon road test (Norway to Portugal)	Mobile station accompanied vehicle	2004	Set world distance record of 6,059 miles
	Three HydroGen3 fuel cell cars	FedEx, METI/JHFC project	Tokyo area		2003–04	1st commercial test of fuel cell vehicle in Japan
Honda	Two 78-kW FCX fuel cell passenger cars		San Francisco		2004	One-year lease with 2nd-year option.
	Two FCX fuel cell passenger cars	South Coast Air Quality Management District	Southern California			
	Two FCX fuel cell passenger cars		New York state; Japan		2005	On-road tests of cold-weather fuel cell stack
Nissan	X-Trail 85-kW SUV	Cosmo Oil	Japan		2004	Lithium-ion battery
Peugeot Citroën	Fuel cell cars	Millennium Cell; Gaz de France	Europe		2004	Hydrogen on Demand fuel system
Toyota	Prius fuel cell hybrid	Texaco Ovonic venture				Demonstrates onboard metal hydride storage
	Highlander SUV hybrid	University of California-Irvine	Southern California	Stuart Energy Station	2004	
Volkswagen	Santana fuel cell sedans	Shanghai Auto Industry	China's Tongju University		2005	Shanghai Auto makes almost half of passenger vehicles sold in China

TABLE 4–2. DEMONSTRATION AND DEVELOPMENT OF HYDROGEN FUELING STATIONS

Oil Company	Vehicles	Partners	Locations	Dates	Notes
BP	Fifteen DaimlerChrysler Citaro urban transit buses	CUTE	Barcelona; Hamburg; London; Porto, Portugal; Stuttgart, Germany	2001–2006	
	Three DaimlerChrysler Citaro urban transit buses		Perth, Australia		Hydrogen supplied by BP's Kwinana refinery
	Hydrogen powered passenger cars		Singapore		
	Hydrogen powered passenger cars		Los Angeles		
	Airport vehicles	Linde, HEW, BMW	Munich		1st public station to offer liquid hydrogen
	Public refueling station for range of alternative fuels		Berlin		
	Up to 30 67-kW Focus passenger cars	Ford; U.S. DOE/ FreedomCAR	Sacramento, CA (10); Orlando, FL, and Detroit, MI	2005	
	Three DaimlerChrysler Citaro urban transit buses	China's Ministry of Science and Technology	Beijing	2005	
	DaimlerChrysler Citaro urban transit buses	China's Ministry of Science and Technology	Shanghai	2005	
Shell	Three fuel cell buses	Ectos/CUTE, Norsk Hydro	Reykjavik, Iceland	opened April 2003	Shell partner in Iceland New Energy Ltd.
	Three fuel cell buses	CUTE; Amsterdam transport company (GVB)	Amsterdam	2003	Test in dense city traffic conditions on normal bus routes
	Three fuel cell buses	CUTE	Luxembourg		Test hilly terrain
		JHFC/METI	Tokyo area		Shell to build and operate one of five stations
	Six HydroGen3 minivans	GM, U.S. Postal Service	Washington, D.C.	2004	U.S.'s 1st hydrogen pump at retail station
	Future residential fuel cell vehicles, HomeFueler	Stuart Energy Systems			

Hydrogen Fueling At Home

Automakers and energy companies are working with manufacturers of hydrogen fueling equipment to develop small, multipurpose systems for homes and businesses. These might eventually pave the way toward individual consumer participation in electricity markets, if interactive connections with utility grids become practical.

Honda/Plug Power

With partner Honda R&D, Plug Power is developing the Home Energy Station—a fuel cell system that uses natural gas to provide electricity and heat in the form of hot water, plus hydrogen for a fuel cell vehicle. The system derives all its energy from the natural gas and requires no grid electricity (Fig. 4–11).[15]

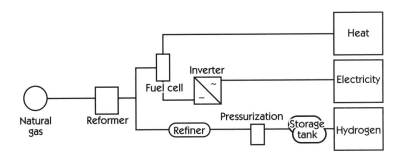

FIG. 4–11. HONDA/PLUG POWER HOME ENERGY STATION. THIS EXPERIMENTAL SYSTEM USES NATURAL GAS TO SUPPLY HYDROGEN TO FUEL CELL VEHICLES AND ELECTRICITY AND HEAT FOR HOUSEHOLD USES. ARTWORK COURTESY OF HONDA.

A prototype at Honda's California research facility demonstrated that it could generate, store, and deliver hydrogen to a Honda fuel cell vehicle and export electricity to the grid. A next-generation design is under way.

The Home Energy Station will supply hydrogen for road tests of Honda's FCX vehicles in California. Throughout the day, the system being tested produces enough hydrogen to refill the tank of a Honda FCX hydrogen car in just a few minutes. The product could enter the market around 2015–2020.

Shell/Stuart Energy

Stuart Energy Systems is a leading supplier of electrolyzers for onsite hydrogen production. With partner Shell, Stuart's European unit is developing the HomeFueler® for vehicles as part of its Personal Energy Station product line. Stuart is also demonstrating a small fueling station that operates directly on solar photovoltaic power with partner Solar Integrated Technologies, a manufacturer of solar canopies.[16]

In a partnership with Toyota, a Stuart Energy Station is being demonstrated in a fuel cell vehicle fueling application in California (Fig. 4–12). Stuart installed Toyota's first hydrogen fueling station and joined Toyota's Fuel Cell Vehicle Community in 2004.

Fig. 4–12. Stuart Energy Station. A Toyota fuel cell vehicle is filled with hydrogen from a Stuart station at Toyota's U.S. headquarters in California. Photo courtesy of Stuart.

Stuart is providing electrolytic hydrogen production for four of Europe's CUTE bus demonstrations. Stuart's compression, storage, and dispensing modules will also be integrated with ChevronTexaco's proprietary reformer technology in a hydrogen fueling station project. Stuart also offers a 125-kW power generation module that uses a hydrogen-fueled internal-combustion engine.

First hydrogen/electricity station

As a forerunner to these small systems, the U.S. DOE funded development of the world's first hydrogen station that provides both vehicle fuel and electricity. The station, which opened in Las Vegas in 2002, is located in the city's vehicle maintenance and operation service center.

The energy station combines an onsite natural gas reformer, a stationary PEM fuel cell, a compressor, liquid and gaseous hydrogen storage tanks, and fuel dispensing systems. In addition to providing hydrogen vehicle fuel, the station can dispense hydrogen-enriched natural gas and compressed natural gas.[17]

The facility was developed by a public-private partnership among DOE, the City of Las Vegas, Plug Power, and industrial hydrogen supplier Air Products and Chemicals, Inc. The $10.8 million demonstration project will operate through 2007.

Utilities Involved in Hydrogen Micropower, Vehicles

Many electric utilities worldwide offer products or services related to micropower. For example, of Japan's ten electric utilities, eight have established subsidiaries to offer micropower services. Installation of distributed power systems grew by 2,418 megawatts during 1997–2000—accounting for 11% of the capacity installed by utilities during the period.[18]

Hydrogen Technology Park

In the U.S. DTE Energy, a Michigan-based electric utility, has built a Hydrogen Technology Park near Detroit where water will be used to produce hydrogen for fuel cell power plants and vehicles. The park, funded by the U.S. DOE, opened in 2004.

The hydrogen energy system at the park is capable of delivering about 100,000 kilowatt-hours of electricity per year—enough to power a small office building or about 20 homes—and enough compressed hydrogen gas to fuel several vehicles per day. Plug Power is supplying the fuel cell equipment, while Stuart Energy Systems is providing the electrolytic hydrogen generator.

The Hydrogen Technology Park will also participate in a DOE project to focus on development of hydrogen fuel cell vehicles and a supporting hydrogen infrastructure. DTE Energy's partners in the project are BP and DaimlerChrysler. At least 20 hydrogen fuel cell vehicles will be tested in Michigan and in California through 2008.

Tech Town microgrid

DTE is also building a microgrid that will be powered by hydrogen, natural gas, and sunlight, which will serve the energy needs of NextEnergy Corp.'s new headquarters building and other facilities at Tech Town, a research and business technology park in Detroit. Microgrids comprise a group of homes or larger buildings that are interconnected and serviced by multiple micro-power units.

DTE's microgrid will use fuel cells, engines (internal and external combustion), turbine technology, and photovoltaic cells. Underground electrical and thermal distribution systems will provide power, heat, and cooling to the buildings. The microgrid construction project should be finished in 2005.

Canada's hydrogen village

A micropower network was under construction in Ontario at the University of Toronto's Missisauga campus as part of the Hydrogen Village Partnership project. Four natural gas-fueled 5-kW solid oxide fuel cells will provide electricity, hot water, and space heating for townhouse-style student residences.

The Hydrogen Village was created by Fuel Cells Canada to promote collaboration among industry, government, and academia. The solid oxide units were supplied by Fuel Cell Technologies Ltd. A series of hydrogen showcases is planned for the greater Toronto area.

Hawaii's plans

Hawaii hopes to demonstrate hydrogen generation from renewable energy sources at a Hydrogen Power Park proposed for the state's Big Island. Design was under way, with funding from the U.S. DOE. If additional funding comes through, the park will install a 75-kW UTC fuel cell and electrolyzers for hydrogen generation. A portable system is planned so that it can be tested on other islands.

Fuel Cell Micropower Technologies

Many businesses could profit from a burgeoning hydrogen economy using fuel cells in vehicles and power plants. Some of the world's biggest manufacturers—including United Technologies and Siemens/Westinghouse—have been working on fuel cell technology for decades, while other players are start-ups led by visionary entrepreneurs.

Most of the stationary fuel cell systems already on the market get their hydrogen fuel from natural gas via a reforming process. Some stationary plants in the future might be powered not by fuel cells alone, but by a hybrid system incorporating a microturbine (see *Microturbines in Stationary Applications* in this chapter).

Fuel cell micropower systems on the market are mostly sized for large buildings and are not commercially available for home use, except for a 1-kW fuel cell unit introduced in Japan in 2004. Many developers are working on residential fuel cells and on improving the economic viability of larger systems. Appendix D lists existing stationary fuel cell applications worldwide.

United Technologies phosphoric acid power plants

UTC began fuel cell research in 1958 and has supplied fuel cells to all of America's 100-plus manned space flights, including the Space Shuttle. Since 1991, UTC Fuel Cells has been selling a 200-kW stationary unit, which it calls the world's first and only commercial fuel cell power plant. In addition to electricity, the PureCell™ 200 produces 900,000 Btu of usable heat. It can be powered by hydrogen, natural gas, or other fuels.[19]

More than 250 PureCell units, which are based on phosphoric acid technology, have been installed in 19 countries on five continents. These applications include a New York City police station, a credit-card processing facility in Nebraska, a science center in Japan, and a high school in Connecticut.

A 1-megawatt fuel cell installation has been operating since 2000 at the U.S. Postal Service facility in Anchorage, Alaska. The system consists of five 200-kW UTC fuel cells using hydrogen derived from natural gas. At the time of startup, this was America's largest fuel cell installation and also the first to be connected to the utility grid. In addition to primary power for the post office, the system provides half of the hot water needed for heating and exports excess electricity to the grid. Local utility Chugach Electric Association installed the system.

UTC is also developing a 5-kW fuel cell power plant for applications such as backup power for telecommunications towers or primary power for small businesses and possibly large homes. Also, UTC's partner Toshiba International Fuel Cells is working on a 1-kW unit for residential applications in Japan.

UTC Fuel Cells also works with the automotive industry, and parent UTC Power has established an alliance with Capstone Turbine Corp. to market Capstone's microturbines for applications that use recovered heat.

Siemens/Westinghouse solid oxide fuel cells

Solid oxide fuel cells are the specialty of this multinational, which has been developing the technology for more than 40 years in its former incarnation as Westinghouse. The company has developed a unique tubular design for solid oxide fuel cells that eliminates the need for seals, and the technology's high-temperature operation allows the cells to reform fuels internally, so no costly reformer is needed (Fig. 4–13).

Commercial prototypes are being manufactured at a pilot facility in Pittsburgh. Siemens/Westinghouse is also testing 25-kW solid oxide units and sells PEM fuel cell modules for submarines and other ships.[20]

Siemens/Westinghouse's demonstration projects include a 100-kW solid oxide fuel cell power plant in Germany that provides hot water for the local district heating system. The company has also developed a hybrid fuel cell/ microturbine system, which produces a total of 220 kW (200 from the fuel cell and 20 from the microturbine). The prototype is being tested by utility Southern California Edison at the University of California–Irvine.

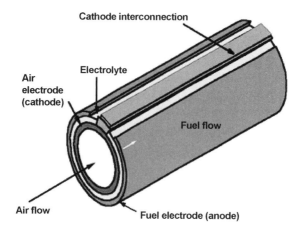

FIG. 4–13. SIEMENS/WESTINGHOUSE SOLID OXIDE FUEL CELL. THIS UNIT
FEATURES A TUBULAR DESIGN THAT ELIMINATES THE NEED FOR SEALS.
ARTWORK COURTESY OF SIEMENS/WESTINGHOUSE.

Siemens/Westinghouse plans to introduce a 250-kW cogenerating solid oxide fuel cell, which is being tested in Canada, and a 500-kW fuel cell-gas turbine hybrid system, based on the experimental 220-kW version.

FuelCell Energy stationary products

This company's fuel cell technology, a different type of solid oxide (planar rather than tubular), uses natural gas or other hydrocarbon fuels directly without extracting the hydrogen separately in an external reformer. Instead, the hydrogen is produced inside the cell and consumed immediately. Fuel-Cell Energy, Inc.'s high-temperature Direct FuelCell® products, which come in sizes of 250 kW to 2 megawatts, were the first to be certified to new U.S. standards for stationary fuel cell packages.[21]

A 250-kilowatt Direct FuelCell was selected as part of a temporary microgrid for the Democratic National Convention in Boston in 2004. The company's micropower plants are also being used in three Sheraton hotels in New York and New Jersey, a Sierra Nevada brewery in California, Yale University's Environmental Science Center in Connecticut, and a municipal wastewater treatment plant in California. FuelCell's Japanese partner Marubeni Corp. operates five units and has ordered several more. Several units have also been installed in Germany and one in Spain.

Automotive fuel cells in an industrial application

The world's largest automaker is joining with the world's largest chemicals manufacturer to propel hydrogen fuel cell technology into the industrial marketplace. Beginning with a single 75-kW automotive PEM fuel cell made by GM, the project will expand to provide Dow Chemical Co. with 35 megawatts of electricity—enough to power 25,000 U.S. homes, but only 2% of consumption at Dow's Texas facility.[22]

Following installation of the first fuel cell in 2004, GM and Dow launched the second phase, a 1-MW integrated pilot plant. The goal is to help drive down the cost of automotive fuel cells while using up some of the huge amounts of hydrogen produced during chemicals manufacturing, which had been used as boiler fuel or sold as a commodity to industrial gas companies.

The 75-kW PEM fuel cells were specified in anticipation of their eventual use in vehicles. The units would be packaged on trailers, each carrying 14 fuel cells (1.05 megawatts per trailer). GM hopes to reduce the cost of the entire vehicle fuel cell powertrain, including traction motors and controls, to $50/kW, or about $3,750 for the 75-kW package. One avenue would be to reduce the amount of platinum catalyst needed.[23]

Other micropower developments

Longer lasting phosphoric acid technology. A phosphoric acid fuel cell is being developed that could run for 7-1/2 years, or about 50% longer than conventional technology. Fuji Electric Holdings Co.'s R&D unit modified a 100-kW phosphoric acid fuel cell with a longer-life stack and reformer—components that previously had to be replaced every five years. Original units are operating in 15 sites. Fuji plans to introduce the longer-life phosphoric acid fuel cell in 2005. The company is also testing a 1-kW residential fuel cell.

Fuel cells using hydrogen from paint fumes. Ford's Dearborn truck plant, which produces the F-150 pickup, is testing fuel cells that use hydrogen derived from paint fumes. The fuel cell system at Ford's Rouge Center converts VOCs—an undesirable by-product of painting vehicles—into hydrogen fuel.

Partners with Ford in the demonstration project are DTE and Canada's Fuel Cell Technologies, which provided the fuel cell system consisting of three 5-kW solid oxide modules. If testing is successful, the fuel cell modules could be applied in other industrial processes that generate VOCs.

Carbonate fuel cell applications to expand. Molten carbonate technology is not as widespread as solid oxide fuel cells, but it shares the features of high-temperature operation and internal fuel reforming. MTU CFC Solutions GmbH, a joint venture formed in 2003, plans to begin producing carbonate fuel cells in 2006. Previously, the company had installed 14 of its carbonate Hot-Modules, mostly in Germany but also in the United States, Spain, and Japan.

MTU CFC plans to market the carbonate technology for fuel cell applications requiring high-temperature heat, such as pharmaceuticals and food processing, and as an emergency power generator. In addition to natural gas, the HotModule can run on biogas, sewer gas, and pit gas.

PEM Fuel Cells in Micropower Applications

Plug Power and Ballard both are advancing PEM fuel cell technology for stationary power, including some models on wheels that could be called portable (though not as portable as a cellphone or laptop).

Residential and commercial micropower

Ballard and partner Ebara have developed a 1-kW PEM fuel cell package that was introduced to the Japanese residential market in 2004 and is still under development for U.S. applications (Fig. 4–14). The fuel cell cogeneration system provides electricity, heat, and hot water to the home and is designed to be connected to the utility grid.

Fig. 4–14. Japanese residential fuel cell. Ebara and Ballard collaborated on developing a 1-kW fuel cell for Japanese homes which provides heat and hot water as well as electricity. Photo ©Ebara.

The Japanese version is fueled with kerosene and incorporates a reformer. According to Ebara, the fuel cell reaches a 35% electrical efficiency, which remains high at partial loads (33% at 50% of rated load), and a heat recovery efficiency of 58%, for a total cogeneration efficiency of 93%. The American version would be fueled with natural gas.

A larger stationary PEM fuel cell system has also been tested by Ballard for several years. The 250-kW unit would produce enough electricity for a small apartment complex, a commercial building, or a group of 50 to 60 houses. Ballard also produces a fuel cell generator on wheels, designed for indoor use (see *Portable applications for fuel cells* in this chapter).

Uninterruptible power supply applications

Ballard also makes 1-kW PEM fuel cell modules designed for replacing dead batteries in standard rack-mounted uninterruptible power supply (UPS) systems. Like a battery, the fuel cells produce d-c voltage that is bused to the UPS. They can operate in parallel up to power levels of 50 kW. Hydrogen is piped to the cell from replaceable containers. Ballard's modules are distributed by Heliocentris Energy Systems.

As an alternative to compressed hydrogen canisters, Texaco Ovonic is developing metal hydride technology for bulk storage of hydrogen. The metal hydride could supply hydrogen to fuel cells in UPS systems and other standby power applications, and could also be used to store hydrogen produced from renewable power sources.

The bulk metal hydride systems can hold 10 kilograms of hydrogen, which is scalable to larger sizes, and deliver the fuel at high pressure (up to 500 psig), eliminating the need for a compressor. A smaller model is available for rack mounting. The storage technology is also being tested in the Toyota Prius fuel cell vehicle.

Millennium Cell's 1.5-kW Hydrogen on Demand system also complements UPS and standby fuel cell applications by providing hydrogen generated from sodium borohydride. By storing more hydrogen than compressed gas canisters taking up the same space, the system can supply fuel to the emergency generator for a longer time.

Plug Power's stationary applications

Plug Power supplies hydrogen PEM micropower plants called GenSys™ in sizes of 2.5 to 5 kW which are designed to run in parallel with the grid but can switch to standby power mode during outages. The company has sited hundreds of fuel cell systems in eight countries and also offers onsite hydrogen generation systems that run on natural gas.

Four of Plug Power's fuel cells have been installed at dairy farms in New York state. A 5-kW unit at Wagner Farms, a 300-cow operation, uses the fuel cell to heat its milking parlor during the winter. The dairy fuel cell systems use LPG rather than natural gas, which is not usually piped to rural locations.

Plug Power's GenCore® hydrogen fuel cell product line is designed for backup power. A 5-kW GenCore system began operating in 2004 to support digital switching gear at a rural telephone network in Canada.

Japanese developers planning small fuel cell systems

Cogeneration systems based on fuel cells, like the one introduced by Ebara and Ballard, are under development for the Japanese market. Tagaki Industrial Co., Ltd., which makes water heaters, will integrate its equipment with a 5-kW PEM fuel cell made by Nuvera Fuel Cells, Inc.

The micropower system will run on natural gas, providing electricity and hot water. The Japanese Gas Association is testing a unit. Field tests in 2005-06 are expected to result in market entry in 2007. Nuvera is also developing a larger micropower system, 75 to 300 kW, based on a proprietary hybrid PEM fuel cell-turbocharger technology. Its electrical efficiency could exceed 45%.[24]

Microturbines in Stationary Applications

Microturbines, as their name implies, are very small turbines based on automotive turbocharger technology with outputs of 25 to 300 kW, although sizes larger than 100 kW are not yet widely available. Microturbines are compact, lightweight, and low-maintenance, with only one moving part—a rotating shaft that turns the turbine, compressor, and generator. Many designs are air-cooled, and some even use air bearings, which eliminates the need for cooling-water and lubricating-oil equipment.

Microturbine shafts spin at up to 100,000 rpm. Shaftpower is converted to a-c electricity through a conventional gearbox and generator. Some microturbines use a high-speed generator that produces d-c electricity, which is converted by power electronics into a-c. Heat from microturbine exhaust gas is recovered by recuperators that transfer the heat energy back into the combustion air stream.

As mentioned earlier, UTC Power plans to market Capstone microturbines in cogeneration applications, and Siemens/Westinghouse is testing a hybrid fuel cell–microturbine system, with plans to introduce it as a 500-kW package. However, Capstone's microturbines and others on the market operate on natural gas or other hydrocarbon fuels and are not yet designed for hydrogen. Microturbine technology is being developed to accept biogases with high concentrations of hydrogen sulfide.

Integrating fuel cells with microturbines

In hybrid applications such as Siemens/Westinghouse's, however, micro-turbines can be powered by the fuel cell's by-product heat. Another example of this application is a Capstone microturbine that was modified to operate on pressure supplied by the high-temperature exhaust heat from a 250-kW solid oxide Direct FuelCell made by FuelCell Energy (Fig. 4–15). The combined electrical outputs give the system high fuel efficiency and low emissions.

FIG. 4–15. CAPSTONE/DIRECT FUELCELL HYBRID SYSTEM. A MODIFIED
CAPSTONE MICROTURBINE OPERATES ON PRESSURE SUPPLIED BY THE HIGH-
TEMPERATURE EXHAUST HEAT FROM A 250-kW SOLID OXIDE DIRECT FUELCELL.
PHOTO COURTESY OF CAPSTONE.

FuelCell Energy is working on this hybrid concept, which uses a network of heat exchangers to transfer waste heat from the Direct FuelCell system to the turbine, which converts a portion of the waste heat to mechanical energy and then electricity. The system could add 10 to 15 percentage points to the efficiency of the fuel cell alone, approaching net electrical efficiencies of 70% to 80%. According to FuelCell Energy, its hybrid concept lets the fuel cell operate at its optimum pressure, independent of the turbine's pressure ratio, allowing the system to be scaled up to larger sizes.

Microturbines in energy systems with fuel cells

In other applications, the fuel cell and microturbine are not integrated but are used in the same micropower system. One example is an installation in Berlin, where two 5-kW PEM fuel cells and a 28-kW natural gas-fueled microturbine will provide electricity, heating, and cooling at a government building.

Partners in the project are IdaTech, which sells PEM fuel cells of 3 to 6 kW, and the fuel cell unit of German utility RWE AG. IdaTech also offers a system that integrates a 1.2-kW fuel cell with a 1-kW solar photovoltaic array.

Other projects using both microturbines and fuel cells include the Tech Town microgrid being built in Detroit by utility DTE.

Fuel Cell Boats, Trains, and Planes

Developments in fuel cell technology have gone beyond automotive applications and are extending into other forms of transportation as well. In San Francisco Bay, a public water taxi ferries passengers using a fuel cell-battery hybrid electric engine. The 18-passenger fuel cell boat was funded by California's Center for the Commercial Deployment of Transportation Technologies, a public-private partnership. The water taxi's Power-X fuel cell technology was provided by Anuvu, which also packages the Nissan Frontier fuel cell pickup truck.

Another fuel cell water taxi, also funded by the California center, was first demonstrated in Newport Beach in 2003. The hydrogen-powered ferry uses a fuel cell provided by Millennium Cell, which is also working with Peugeot on fuel cell vehicles.

The German Navy is building four U212 class, fuel cell-powered sub-marines. The first one, launched in 2002, uses four PEM fuel cells, each pro-viding 30 to 50 kW, for slow, silent cruising with no telltale heat emissions. The sub also contains a conventional diesel generator. Italy launched its first fuel cell submarine in 2003.

Locomotive being developed

An international consortium is developing the world's largest fuel cell vehicle, a 1.2-megawatt locomotive for defense and commercial railway appli-cations. The behemoth weighs some 120 tons. The five-year project is led by Vehicle Projects LLC and funded by the U.S. Army.

Design of the fuel cell locomotive was completed in 2004, including onboard fuel storage, offboard hydrogen generation plant, refueling system, and fuel cell power plant. The locomotive will use ammonia-based hydrogen supply, metal hydride storage, and PEM fuel cell stacks.

A smaller locomotive—150-kW, 25 tons—was also being developed for underground mining applications, where safety and health regulations limit vehicle exhaust. A 4-ton prototype has been tested aboveground.

The sky's the limit?

Hydrogen-fueled airplanes have taken off in the past, beginning in the 1950s with jet bombers powered partially by liquid hydrogen. Liquid hydrogen contains about 2.8 times as much energy as jet kerosene on a pound-for-pound basis, but takes up three to four times as much space. So an airplane's hydrogen fuel tank would be lighter but bigger than a kerosene tank. Although most air pollutant emissions would be reduced, a hydrogen-fueled airplane would produce nitrogen oxides.[25]

Recently, little progress has been made in hydrogen-powered airplanes for civilian use. NASA evaluated the concept of a hydrogen fuel cell aircraft and found it to be impractical without dramatic technology advances. The propulsion system would be much heavier, and the plane would probably be noisier than conventional jets.

However, Aviation Tomorrow is building a small plane that could achieve hydrogen-powered, piloted flight if tests go well. The prototype aircraft, called the E-plane (Fig. 4–16) and based on the Diamond Motorglider airframe, is being equipped with a 10–15 kW fuel cell affording a 500-mile (800-km) travel range, plus emergency battery power.

FIG. 4–16. E-PLANE PROTOTYPE. AVIATION TOMORROW HOPES TO EQUIP THIS E-PLANE WITH A FUEL CELL, OFFERING HYDROGEN-POWERED, PILOTED FLIGHT.

Portable Applications For Fuel Cells

Fuel cells can be used to power a variety of portable devices, from handheld electronics such as cellphones and boomboxes to larger equipment such as portable generators. Potential applications include laptop computers, digital cameras and camcorders, and personal organizers—almost any device that uses batteries. Worldwide, an estimated 1,700 portable fuel cell systems have been developed and operated. Market analysts expect portable applications to enjoy widespread market success sooner than automotive or stationary fuel cells.

The primary technology using hydrogen for portable applications is the PEM fuel cell. However, some manufacturers are working on propane-fueled solid oxide fuel cells, and many of the biggest players favor direct-methanol designs, where hydrogen is not extracted for fuel, and the amounts used are so tiny that emissions become less of a factor. Portable fuel cells range in electrical output from 1 Watt for consumer electronics applications to 1.5 kilowatts or more for portable generators and remote power units.

Competing with batteries

Today's wireless revolution is being restrained by battery life. Most of us have discovered at one time or another that our cellphone or digital camera has lost its charge. In portable applications, fuel cells could last much longer than batteries before refueling is required, and they should soon compete on cost as well. Christopher Hebling of Germany's Fraunhofer Institute believes that mass-produced hydrogen fuel cells could be cheaper than rechargeable lithium-ion batteries within a few years.[26]

In the smallest fuel cell devices, a cartridge of metal hydride similar to an inkjet printer cartridge might be used to refuel the device. Larger equipment could use canisters of compressed hydrogen, metal hydrides, or other advanced storage technologies.

The market for laptop computer batteries alone was over $1 billion per year in 2004, said Business Communications. By 2012, portable fuel cells are expected to power 13.5% of the world's laptops. Market entry should begin in 2005, along with fuel cell applications in personal digital assistants (PDAs). Global shipments of portable fuel cells could reach 120 million units by 2012. However, mass market acceptance could be restrained if refueling or replacing fuel cartridges is too expensive.[27]

Early market entries

A prototype PEM cell producing 50 Watts has been developed by Germany's Masterflex AG and Fraunhofer Institute. The fuel cell stores hydrogen in a miniature metal hydride system that holds 15 times the capacity of a conventional battery, allowing the unit to power a notebook computer for up to 35 hours. Masterflex plans to market the PEM fuel cell in 2005.[28]

The two partners also introduced their Mobile Power Box in 2003, a hydrogen fuel cell package designed to supply off-grid power up to 100 Watts. The system is targeted at portable applications such as traffic signals and roadwork signs.

Another PEM fuel cell has been developed for laptop computers that extracts hydrogen from methanol. Manufacturer Casio plans to introduce the fuel cell in 2007. The fuel cell is about the same size as a lithium-ion battery, but packs almost four times as much power, allowing it to run a typical laptop for 8 to 16 hours. Direct-methanol versions would require miniature pumps, making them too large for laptops, said Casio.

Personal power

Ballard manufactures a fuel cell generator on wheels, designed for indoor use (Fig. 4–17). The 1-kW AirGen™ unit functions as a portable power generator, allowing users to plug computers and other appliances directly into the system. It can also serve as a source of emergency power for up to 15 hours, protecting home equipment from crashing by starting up automatically during an outage.

Fig. 4–17. Ballard AirGen portable fuel cell generator. This unit can be used for portable power indoors or as an emergency generator. Photo courtesy of Ballard Power Systems.

Unlike battery backup systems, it continues running as long as hydrogen fuel is supplied. Hydrogen fuel canisters snap into place and can be replaced while the unit is operating. Heliocentris distributes the AirGen systems. Ballard also makes hydrogen fuel cell modules generating up to 1200 Watts of d-c power for integration in portable products.

Solid oxide portable fuel cells

A 50-Watt solid oxide fuel cell was introduced in 2004 that runs on propane but can be adapted to natural gas or other hydrocarbon fuels. As with larger solid oxide systems, the fuel is reformed into hydrogen internally. NDEenergy was marketing the fuel cell for portable power and planned to introduce a 100-Watt model.

As with aerospace fuel cell technology, the military has been a developer and early adopter of fuel cells for portable use. A 10-kilowatt portable generator powered by a solid oxide fuel cell is being developed by Gas Technology Institute for the U.S. Army. The fuel cell would use diesel or other available military fuel.

Fuel cell shopping online

DCH Technology and FuelCellStore.com host an Internet website (www.dcht.com) that sells fuel cell systems directly to customers, but with hefty price tags. Products include portable hydrogen fuel cells ranging from 100 to 200 Watts, with metal hydride or compressed gas storage canisters sold separately.

The online shop also offers portable fuel cell generators such as Ballard's AirGen. You can even buy a 60-watt radio-controlled fuel cell car, if you have $6,600 in spare change. If not, you can order a model fuel cell car kit from uToypia for about $120—water not included.

Consumer Demand

The future of a hydrogen economy based on fuel cells might depend not on replacing our existing energy-powered equipment, but on consumer demand for new products and services that we can't imagine today. Novel uses for hydrogen could add value to peoples' everyday lives, perhaps just by saving time.

Convenience and usefulness could drive demand and accelerate mass production, said Arno A. Evers of Fair-Pr. History shows that completely new services enter the public market as luxury goods at a very high price, affordable only by elite, wealthy consumers. If the service or product offers desirable advantages, demand will grow and production costs will fall, lowering prices. The breakthrough for everyday fuel cell applications could come from a new service with unforeseen benefits.[29]

One example might be what Evers calls the personal power car, similar to Amory Lovins's concept described in chapter 1, which would be fueled initially by reformed natural gas and eventually by renewable power. However, consumers would have to be able to save money by using heat and electricity produced by the car and also make money by selling power from it while the car is parked.

In the future novel, non-energy applications for hydrogen might emerge, such as its potential use in the fight against cancer. Early experiments have suggested that breathing hydrox—pressurized hydrogen mixed with a safe amount of oxygen—might have beneficial effects on tumors. The theory is that hydrogen (a chemical reducing agent) might serve as an antioxidant, scavenging the free hydroxyl radicals that are known to damage cells.[30]

The concept of producing food from hydrogen has also been explored for decades. The idea would be to create single-cell proteins from hydrogen, using bacteria. Protein can already be produced from crude oil and natural gas. Scientists are still studying microbes that can produce protein from hydrogen and other inorganic materials, using carbon dioxide as the carbon source. Bacterial protein produced in this way is tasteless and dissolves easily, similar to egg white or tofu.

References

1. Parrish, Alton, "Hydrogen Market to Nearly Double to US $1.5 Billion By 2008," *Fuel Cell Technology News*, January 19, 2004.

2. "U.S. merchant and on-site hydrogen sales to reach $2.7 billion by 2008," *Power Engineering*, October 13, 2003.

3. Electric Power Research Institute, "Distributed Energy Resources Market is Starting to Rebound, Says Primen: New Study Identifies Key Factors Influencing Purchase Decisions," January 26, 2004, www.epri.com.

4. Lord Browne in Vaitheeswaran, Vijay V., *Power to the People: How the Coming Energy Revolution Will Transform an Industry, Change Our Lives, and Maybe Even Save the Planet*. Farrar, Straus and Giroux: New York, 2003. ISBN 0-374-23675-5, page 218.

5. Mogford, John, "Growing Markets for Cleaner Energy: The Role of Hydrogen within an Energy Major," Hyforum Conference, Beijing, May 25, 2004, www.bp.com.

6. EERE Network News, U.S. Department of Energy, Office of Energy Efficiency and Renewable Energy, July 7, 2004, www.eere.energy.gov.

7. General Motors Corp., www.gm.com.

8. Ford Motor Co., www.ford.com.

9. Glover, Mark, "Hydrogen-powered Fords set for test-drive in capital," *Sacramento Bee*, April 28, 2004, www.sacbee.com.

10. Vaitheeswaran, page 259.

11. Hoffmann, Peter, "The Hydrogen Power Rush," 2002, www.worldandi.com.

12. Fuel Cell Bus Club, www.fuel-cell-bus-club.com.

13. Japan Hydrogen & Fuel Cell Demonstration Project, www.jhfc.jp.

14. California Hydrogen Highway, www.hydrogenhighway.ca.gov.

15. "Plug Power and Honda demonstrate Home Energy Station," *Fuel Cell Today*, October 2, 2003, www.fuelcelltoday.com.

16. Stuart Energy Systems Corp., www.stuartenergy.com.

17. Blankinship, Steve, "When Will the Hydrogen Future Arrive?", *Power Engineering*, January 2003, http://pe.pennwellnet.com.

18. International Energy Agency, "Distributed-Generation Issues in Japan, the US, the Netherlands, and the UK," www.iea.org.

19. United Technologies Corp., UTC Power Division, UTC Fuel Cells, www.utcfuelcells.com.

20. Siemens/Westinghouse, www.siemens.com.

21. FuelCell Energy, Inc., www.eerc.com.

22. Dow Chemical Co., www.dow.com.

23. Brooks, Bob, "GM, Dow to Begin Fuel-Cell Power-Generation Project," July 8, 2003, WardsAuto.com.

24. Nuvera Fuel Cells, www.nuvera.com.

25. Hoffman, Peter, *Tomorrow's Energy: Hydrogen, Fuel Cells, and the Prospects for a Cleaner Planet.* The MIT Press: Cambridge, MA, 2001. ISBN 0-262-08295-0, pages 165–168.

26. Labs, Wayne, "Fuel Cells Go Back to the Future—Again," *Electronic Design,* June 14, 2004, pages 83–85.

27. Business Communications Co., "10–15% of Laptops Will Run on Micro Fuel Cells in 2012, Forecasts ABI Research," May 12, 2004.

28. Masterflex, www.masterflex.de; Labs, Wayne, pages 83–85.

29. Evers, Arno A., "Why Should I Buy a Fuel Cell?", February 7, 2003, www.energypulse.net.

30. Hoffman, Peter, pages 222–228.

How would a Hydrogen Economy work?

A future hydrogen economy would require the production, transport, storage, and distribution of hydrogen at reasonable costs, with moderate carbon emissions, and with acceptable levels of safety. We need to improve our methods of producing and storing hydrogen, apply new technologies to capture and sequester emissions of carbon dioxide, and develop cheaper ways to deliver hydrogen to customers or produce it onsite, at the point of use.

Hydrogen isn't found free in nature, but it can be produced from water, hydrocarbon fuels, and other compounds. Most of our hydrogen supplies in the near term would come from reforming natural gas in existing central plants and in small, onsite units at hydrogen vehicle fueling stations. Electrolysis systems that split the water molecule will also supply onsite hydrogen in the near to mid-term. Reforming systems and electrolytic units are already operating at dozens of fueling stations worldwide and are being scaled down for residential use.

In the future, more hydrogen would be produced via electrolysis using renewable power sources such as wind turbines and solar energy. Biomass could also provide large quantities of hydrogen if conversion processes become more efficient. The use of coal to produce hydrogen would depend on cost-effective techniques to sequester carbon dioxide. Future technologies for making hydrogen might also include high-temperature water-splitting using nuclear power or concentrated solar heat, as well as advanced biological techniques using genetically engineered algae.

Today's hydrogen storage technologies—compressed gas cylinders and liquid hydrogen tanks—are inadequate for cost-effective use onboard fuel cell vehicles. The automotive industry's requirement for mass-market success is a travel range of 300 miles (500 kilometers) without an unreasonable weight penalty and without taking up too much valuable cargo space. Existing storage systems are being improved, and new technologies are being developed to store hydrogen in compounds called hydrides and in other materials such as microscopic tubes of carbon.

Hydrogen safety practices are well-established in several industries and would be more widely applied in a future hydrogen economy. Like gasoline, hydrogen can pose safety risks in enclosed structures and during vehicle refueling, but hydrogen can be as safe as any other fuel with proper handling and controls. Codes and standards are being further developed to ensure hydrogen safety, but an educational program would be needed to raise awareness among local officials, rescue workers, and the general public.

Existing methods of transporting compressed or liquefied hydrogen by truck are much too costly for widespread application, and construction of an extensive hydrogen pipeline network would be too expensive to consider in the near future. During a transition toward a hydrogen economy, some deliveries by truck would fill gaps in the supply chain, but most hydrogen would be produced onsite to avoid high transportation costs. Eventually some dedicated hydrogen pipelines might be built near large complexes of refineries and chemical plants, which already produce and consume huge amounts of hydrogen.

Building a network of hydrogen fueling stations will be critical to the success of hydrogen as a transportation fuel, due to consumer demand for convenience. Estimates differ on how much this infrastructure would cost, but automakers and oil companies think that an initial system of fueling stations could be built for a reasonable investment (less than $20 billion in the United States). The minimum requirements for consumer convenience would be hydrogen availability within 2 miles in cities and within 25 miles cross-country (3 and 40 kilometers, respectively).

These topics are discussed in chapters 5, 6, and 7.

Production and **Supply**

One of hydrogen's advantages is that it is found in a great variety of compounds, including water, hydrocarbon fuels, and inorganic substances. Like electricity, hydrogen can be produced from a multitude of sources and in many different ways. This flexibility of production would allow hydrogen to be phased into our existing energy systems in the least disruptive way. If new sources of energy are discovered and developed in the future, they could also be used to make hydrogen, without affecting the way that energy services are delivered to customers.

However, unlocking the chemical bonds in hydrogen-containing compounds requires energy—electricity to split water into hydrogen and oxygen, for example, or heat to extract hydrogen from fossil fuels such as natural gas. The cost of producing hydrogen by these conventional methods or by other, more advanced technologies is one of the biggest barriers on the path toward a hydrogen economy.

Another concern is the environmental impact of using fossil fuels as feedstocks or power sources for hydrogen production. Hydrogen's potential to reduce global emissions of greenhouse gases (primarily carbon dioxide, or CO_2) would be partially negated by the carbon released when hydrocarbon feedstocks are processed into hydrogen. Similarly, fossil fuels still generate a large share of most industrial nations' power supplies, so splitting water with this electricity to produce hydrogen also would add to CO_2 emissions.

During a transition toward a hydrogen economy, these carbon emissions would have to be captured and stored away (sequestered) until enough hydrogen can be produced from clean, renewable sources of energy.

In this chapter, we'll explore several hydrogen production methods and take a look at the costs, environmental concerns, and other barriers that would have to be overcome for widespread use. Ultimately, the goal is for hydrogen to be produced at industrial facilities, power parks and microgrids, local fueling stations, and customers' homes and businesses. Processes could use locally available fossil fuels or biomass feedstocks, releasing little or no CO_2, and water would become just another feedstock to be electrolyzed by renewable power.

Overview of Production Methods

Huge amounts of hydrogen are already produced and consumed at oil refineries, chemicals manufacturing plants, and other industrial operations around the world (see chapter 3). The two primary conventional methods of producing hydrogen today are reforming of hydrocarbon fossil fuels and electrolysis of water. The least expensive, most common process in the United States is catalytic steam reforming of natural gas (mostly methane), while some European nations use nuclear power for electrolysis.

Electrolysis, or water-splitting, is more energy-intensive than reforming and thus more costly in countries where electricity is expensive. The process essentially is the mirror image of a fuel cell, which combines hydrogen and oxygen into water. Although electrolysis is well-proven, it is economical only in areas with cheap electricity or in applications where very pure hydrogen is required.

Several advanced technologies are being developed to produce hydrogen in the future. These include gasification of biomass and coal, photo-electrochemical water-splitting using sunlight, biological methods that produce hydrogen from algae and bacteria, and high-temperature water-splitting using chemicals and solar or nuclear heat.

Units of measurement

Although most industrial hydrogen is measured by volume—standard cubic feet (scf) or normal cubic meters (nm^3)—hydrogen for vehicle fueling can also be measured by weight (kilograms) or gge. A kilogram of hydrogen is the energy equivalent of about a gallon of gasoline.

Production of about 25 kilograms of hydrogen per day (kg/d), or roughly 10,000 kilograms per year (kg/yr), is enough to fuel a small fleet of 5 to 50 vehicles. A full-size fueling station serving more than 500 vehicles would need to produce about 1,000 kg/d, or more than 300,000 kg/yr.

Carbon emissions are measured in kilograms of carbon produced per kilogram of hydrogen produced. The cost of storing carbon is estimated at $37 per metric ton, which translates to about $10 per metric ton of CO_2, although some projections are higher. Assuming carbon is taxed in industrial nations in the future, the estimated cost of carbon emissions is $50 per metric ton. When carbon emissions trading programs are created and expanded, a value per ton would emerge in the marketplace.

The cost of carbon sequestration is fairly small, in the context of conventional energy prices. Carbon storage would represent an additional 9¢ in the price of a gallon of gasoline, $4 per barrel of oil, 54¢ per thousand cubic feet of natural gas, 0.8¢/kWh of coal-generated electricity, or 0.4¢/kWh of natural gas-generated electricity.[1]

Challenges and outlook

Selecting feedstocks for hydrogen production involves economic and environmental tradeoffs. For example, fossil fuel feedstocks are well-understood and accepted, but until carbon sequestration becomes practical, emissions would limit their use. Biomass, on the other hand, has very small net carbon emissions, but we'd have to figure out how to grow, collect, transport, and store the massive amounts required.

Today large, central steam methane reformers produce most of the hydrogen in America and could continue to meet hydrogen demand in the near term. However, smaller reformers have been developed to produce hydrogen onsite, at the point of use. Many of these systems are now being demonstrated in hydrogen fueling stations for fuel cell vehicle fleets, and at least one is being scaled down to residential size for integration into a Home Energy Station (see chapter 4).

As hydrogen use increases and hydrogen markets grow, small onsite reformers would proliferate, central reforming plants would incorporate carbon sequestration to reduce emissions, and nuclear plants globally could produce increasing amounts of hydrogen using new technologies.

Other near-term hydrogen production methods, such as conversion of biomass to usable liquids and gases, will require improvements in system efficiency and integration to compete with hydrogen produced from methane reforming.

The capital costs of electrolysis systems, along with high electricity costs and carbon emissions from fossil-fueled power plants, could limit the adoption of electrolyzers for producing hydrogen. Research is under way to reduce the cost of electrolyzers and improve their efficiency, and electrolysis is being integrated into systems with renewable power sources to provide pollution-free hydrogen.

Many of the technologies with the greatest long-term promise are just emerging and will need financial support to reach their full potential. Photobiological hydrogen production requires technological advances, beginning with genetic engineering of micro-organisms that can produce hydrogen at high rates. Photoelectrochemical processes might depend on a breakthrough in materials science to become practical. And high-temperature hydrogen production using water-splitting chemical cycles requires more work on solar heat concentration and new nuclear reactor designs.

Reforming Natural Gas and Liquids

This process is poised to be the near-term choice for producing hydrogen until cleaner processes using renewable energy become practical. Natural gas reforming has the advantages of high efficiency and low cost, as well as the option of recovering and using waste heat in a cogeneration system.

One downside is dependence on natural gas fuel. Gas prices fluctuate widely, and most industrial nations import much of their supply (see chapter 2). However, natural gas is widely available and easy to handle, and it has the highest hydrogen-to-carbon ratio of any fossil fuel, so it produces relatively moderate CO_2 emissions. In addition to natural gas, a variety of liquid fuels also can be reformed to produce hydrogen.

In the first step of reforming natural gas, methane (CH_4) reacts with steam at temperatures of about 800°–900°C (1,500°–1,600°F), accelerated by a nickel catalyst. This reforming step produces a synthesis gas, or syngas, composed of hydrogen (typically 75%), carbon monoxide (15%), and CO_2 (10%).

Then a reaction with additional steam (called the water-gas shift reaction) converts the carbon monoxide to CO_2 and produces more hydrogen, which can be separated from the syngas and purified. CO_2 is removed by an adsorption process, typically pressure swing adsorption (PSA). Syngas can also be used for many purposes including power generation and oil refining. Overall, the methane steam reforming process (Fig. 5–1) can be represented by:

$$CH_4 + 2H_2O \rightarrow 4H_2 + CO_2$$

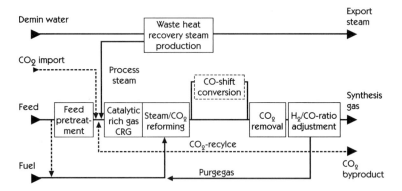

FIG. 5–1. STEAM REFORMING PROCESS. CATALYTIC STEAM REFORMING OF
METHANE AND OTHER HYDROCARBONS PRODUCES A SYNTHESIS GAS FROM WHICH
HYDROGEN IS SEPARATED AND PURIFIED. ARTWORK COURTESY OF LINDE.

Two variations of catalytic steam reforming are in use—partial oxidation and autothermal reforming. In partial oxidation (or partial combustion), some of the feedstock is burned as fuel to generate the heat required for subsequent reactions, but combustion is quenched before it is complete. Oxygen for burning the fuel is provided by cryogenic separation from air, an expensive part of the system. Partial oxidation is better suited for reforming heavier hydrocarbon fuels such as diesel and gasoline, because the initial high-temperature combustion step breaks these large molecules into structures that are simpler to process.[2]

Autothermal reformers combine steam reforming and partial oxidation. Steam reforming requires heat, while partial oxidation gives off heat. By combining the two, the autothermal process becomes only slightly hot overall. This results in a system that is smaller, quicker starting, and quicker responding than a steam reformer without the high operating temperatures of a partial oxidation reformer. Many of the newest onsite reformers use autothermal technology, such as Plug Power's GenSite™ (Fig. 5–2).

Fig. 5–2. Autothermal reformer. Plug Power's GenSite™ onsite
reformer uses autothermal technology, which combines steam
reforming and partial oxidation. Photo courtesy of Plug Power.

Small onsite reformers are also called fuel processors, and some were originally designed for applications onboard vehicles or in micropower plants. (The term hydrogen generator usually refers to an electrolysis-based unit.) However, small reformers or processors generally are not as efficient as large, central reforming plants and emit more CO_2 per unit of hydrogen produced.

In addition to reforming natural gas, many of the units on the market or under development can also process other petroleum-derived gases and liquids, including propane, naphtha, ammonia, methanol, ethanol, and even gasoline. Although natural gas will likely remain the reforming fuel of choice, fuel flexibility is one advantage of onsite reforming as a hydrogen production technology. Fuel-flexible reformers might be able to exploit feedstock price fluctuations and operate on other fuels during shortages in primary fuel supply.

Barriers and research needs

The biggest challenge to widespread use of onsite reforming systems is to mass-produce units that will operate reliably and safely with little attention from relatively unskilled workers (fueling station attendants and consumers), according to the U.S. NRC in its report, *The Hydrogen Economy: Opportunities, Costs, Barriers, and R&D Needs.*[3]

During a transition toward hydrogen energy systems, thousands of small reformers would be required to meet growing demand. New designs will be needed to minimize manufacturing costs and optimize efficiency.

Also, these units must incorporate the ancillary equipment needed to meet varying levels of demand throughout the day. These systems include controls, adjustable output (turndown), on-off cycling, and hydrogen compression and storage. With today's technology, this equipment costs about 30% as much as the reformer itself. With technological improvements, especially in compression and storage, these costs could be cut in half while increasing overall system efficiency.

Other improvements could include better catalysts for both the reforming step and the shift reaction. Existing catalysts are not sufficiently durable, efficient, or tolerant of impurities. Hydrogen separation and purification processes must become more efficient and cost-effective, primarily through development of advanced membrane technology. Also, costs might be reduced by integrating steps in the reforming-shift-separation-purification process.

Partial oxidation and autothermal reformers would need better, cheaper ways to extract oxygen from air. Membrane separation technology is under development for application in these reforming units. Scientists are working on a nonporous ceramic membrane (called an advanced ion transport membrane) that would simultaneously separate oxygen from air and partially oxidize methane.

Hydrogen production economics

With further development, the overall thermal efficiencies of onsite re-formers could increase from 55.5% to 65.2%, said NRC, and the capital cost of a unit producing 480 kg/d of hydrogen (a medium-size vehicle fueling operation) could be reduced by almost 50%, from $3,847 per daily kilogram produced to $2,000/kg-d.

As a result, hydrogen costs from onsite reformers could fall by about 34%, from $3.51/kg to $2.33/kg. NRC's estimates are based on natural gas prices of $6.50/million Btu. Price variations of plus-or-minus $2/million Btu would change the hydrogen cost by about 12% with current technology (the change is not proportional because gas price is only one of the factors in calculating hydrogen costs). Additional information on hydrogen costs is provided in Appendix E.

Carbon emissions

In today's large, central steam methane reforming plants, about 2.5 times as much carbon is emitted per unit weight of hydrogen produced. With carbon capture and storage, this ratio could decline to 0.42 in plants using existing technology, said NRC. For central plants using tomorrow's improved technologies, the estimated ratios are 2.39 without carbon capture and 0.35 with carbon capture and storage.

Hydrogen from natural gas reforming in the future would have to be ex-tracted at relatively high purity for use in fuel cells. However, exhaust streams can contain not only CO_2 but also nitrogen that dilutes the CO_2, making it more difficult to capture. Tomorrow's reforming plants would require ad-ditional equipment and processing not used in existing systems, and new carbon capture technology might be needed.

Despite the added complexity, however, the cost of CO_2 capture would increase the cost of the hydrogen produced by only 8% to 15% (Table 5–1).

TABLE 5–1. HYDROGEN COSTS FOR CENTRAL STEAM METHANE REFORMING

Technology	Hydrogen cost, $/kg H_2						
	Without carbon capture			With carbon capture and storage			
	Plant	Carbon emissions*	Total	Plant	Storage**	Carbon emissions*	Total
Current	1.03	0.13	1.16	1.22	0.09	0.02	1.33
Future	0.92	0.12	1.04	1.02	0.08	0.02	1.12

* At carbon emissions costs of $50 per metric ton.
** At carbon storage costs of $37 per metric ton.
Source: NRC.

Carbon capture from onsite reformers will probably be impractical, since the cost of sequestration appears prohibitive on a small scale. However, technical improvements could mitigate the amount of CO_2 released by about 15%, from 12.1 kg of CO_2 per kg of hydrogen to 10.3 kg/kg.

Vehicle emissions

NRC also estimated the effect on carbon emissions of shifting gasoline vehicles to hydrogen fuel produced via reforming. For a U.S. fleet of passenger cars and light-duty trucks (hydrogen fuel cell vehicles vs. gasoline-electric hybrids), CO_2 emissions would be reduced by 30% using hydrogen from small onsite reforming systems and by 50% from central reforming plants, both without carbon sequestration.

Carbon emissions from hydrogen production in the United States totaled close to 400 million metric tons per year (tpy) in 2000. With continuing use of gasoline light-duty vehicles, carbon emissions would rise to more than 700 million tpy by 2050, said NRC. Switching to fuel cell vehicles running on hydrogen produced by either onsite or central natural gas reforming, even without carbon capture, would reduce carbon emissions in 2050 by more than 400 million tpy.

Carbon Capture and Storage

Capturing carbon emissions from natural gas reforming processes would further improve the environmental advantages of a hydrogen economy. The use of coal, however, would require carbon sequestration to gain any meaningful pollution benefits, whether the coal is used as a hydrogen feedstock or as a power source for splitting water. Although all fossil fuels emit carbon when burned, coal's carbon-to-hydrogen ratio is the highest of the primary fuels.

Although CO_2 is routinely separated and captured in many industrial processes, these techniques would not be cost-effective on the scale required for a pervasive hydrogen economy. Evolutionary improvements in existing systems and revolutionary new capture and sequestration concepts will be needed to bring costs down. The most likely candidate processes for separating and capturing CO_2 are absorption, adsorption, low-temperature distillation, gas separation membranes, and mineralization.

Sequestration technologies

Once captured, carbon dioxide can be stored in underground reservoirs, absorbed in oceans or injected deep underwater, or converted to rock-like solid materials. In geological sequestration, CO_2 would be stored in underground formations such as depleted oil and gas reservoirs, unmineable coal deposits, and salt caverns.

The energy industry already uses CO_2 to pump more oil out of mature fields. When injected underground, the CO_2 increases the oil reservoir's pressure, making it easier to force the remaining oil out. Possibly, CO_2 might perform a similar function by displacing methane trapped in coal seams, allowing the valuable gas to be recovered.

Norway's Statoil has been injecting and sequestering CO_2 underground since 1996. The CO_2 is a constituent of the natural gas produced from the Sleipner field in the North Sea (Fig. 5–3). Norway's tax of about $140/ton of carbon emissions makes the project economical. Norway also plans to reinject CO_2 produced at its Snøhvit gas field in the Barents Sea.

FIG. 5–3. SLEIPNER FIELD. CARBON DIOXIDE SEPARATED FROM NATURAL GAS PRODUCED ON SLEIPNER'S T PLATFORM IS REINJECTED FROM THE A PLATFORM INTO A SUBSEA FORMATION. PHOTO COURTESY OF STATOIL.

BP is planning one of the world's largest CO_2 storage projects at the In Salah gas field in Algeria, where about 1.1 million tpy will be sequestered. BP is also working with Ford through the Carbon Mitigation Initiative at Princeton University, which links fundamental climate science research with practical approaches.[4]

Another sequestration project is under way in Canada, where CO_2 from a coal gasification plant in the United States is piped over the border and injected in the Weyburn oil fields for enhanced oil recovery and CO_2 storage. In the U.S. oil industry, a total of about 25 million tpy of CO_2 is injected for enhanced recovery (Fig. 5–4). Researchers are studying the geology and other conditions at Weyburn and at U.S. dome sites (including McElmo and Jackson) to ascertain how long the CO_2 will remain underground.[5]

FIG. 5–4. U.S. CO_2 INJECTION SITES. THE U.S. OIL INDUSTRY INJECTS ABOUT 25 MILLION TPY OF CO_2 FOR ENHANCED OIL RECOVERY. ARTWORK COURTESY OF *Oil & Gas Journal.*

The United States also contains a wealth of deep salt formations (Fig. 5–5) that could sequester an estimated 500 billion metric tons of CO_2. Although these sites would be less cost-effective because no oil would be produced, researchers project that the formations would hold the CO_2 for a few thousand years.

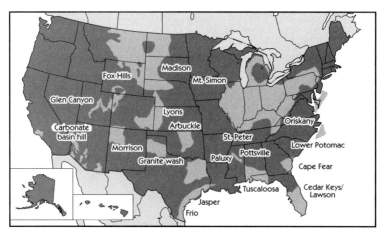

FIG. 5–5. U.S. DEEP SALT FORMATIONS. THESE GEOLOGICAL FORMATIONS COULD SEQUESTER AN ESTIMATED 500 BILLION METRIC TONS OF CO_2. ARTWORK COURTESY OF *Oil & Gas Journal*.

The concept of ocean sequestration is at a much earlier stage of development. Although the world's oceans naturally absorb and release CO_2, they do it very slowly. Scientists are working on enhancing the absorption rate and studying the potential effects, such as increasing the water's acidity, which could upset the ocean's ecological balance. Technologies are also being explored to inject CO_2 directly into the ocean depths.

Sequestering carbon on land in terrestrial ecosystems is also being studied. Like the oceans, our terrestrial biosphere already stores huge amounts of carbon. Researchers are looking at ways to protect certain ecosystems and enhance their ability to absorb CO_2. These include forests, agricultural land, biomass crops, deserts, other wastelands, and wetlands.

Two promising concepts for storing CO_2 in solid compounds are magnesium carbonate and CO_2 clathrate, an ice-like material. Both provide quantum increases in density compared to gaseous CO_2. Biologists are also looking at ways to boost the CO_2 uptake of photosynthetic plants and to harness natural, non-photosynthetic microbial processes capable of converting CO_2

into useful products such as methane. An important advantage of biological systems is that the CO_2 need not be pure.

Before carbon capture and storage is needed for hydrogen production on a large scale, the technologies will probably be demonstrated extensively in other applications such as natural gas purification, power plant emission control, and ammonia production, said NRC.

Government/industry programs

Carbon Sequestration Leadership Forum. This international group comprises 16 nations, including the United States, Japan, China, Russia, and several European members. The forum aims to develop better carbon capture and storage technologies and make them available globally, as well as foster international cooperation and promote policies and regulations that support carbon sequestration.

CO_2 Capture Project. The energy industry, including several leading multinationals, has formed a group that is working to reduce the cost of CO_2 capture from combustion sources and develop methods for storing CO_2 underground. Members include BP and Shell, as well as ChevronTexaco, Eni SpA, Statoil, Norsk Hydro, EnCana, and Suncor Energy. The companies work with environmental advocates such as the Natural Resources Defense Council. Technologies being developed include CO_2 capture from natural gas reforming.

Environmental concerns

Although some environmentalists have adopted a wait-and-see attitude toward CO_2 capture, many others—including some supporters of hydrogen fuel cells—are uncomfortable with the idea of making hydrogen from fossil fuels. As Peter Hoffmann explains in his book *Tomorrow's Energy*, "They see it as backsliding under the guise of logistical concerns in a Faustian pact that negates the entire reason for going to hydrogen in the first place."[6]

Another pathway to hydrogen is also unpopular with the green community—using nuclear power to split water. Hydrogen produced from nuclear power or from coal is sometimes called black hydrogen, as opposed to green hydrogen produced from clean, renewable power sources. Many environmental advocates see disturbing similarities in the idea of storing both carbon and nuclear waste underground or in the oceans.

Coal is the more immediately troublesome environmental concern because the United States and fast-developing nations such as China contain such huge reserves of the fossil fuel, and because of its high carbon content and pollution potential.

Producing Hydrogen From Coal

The primary route for producing hydrogen from coal is via gasification. Rather than burning coal directly, gasification breaks coal down into its basic chemical constituents. The coal is exposed to steam, along with carefully controlled amounts of air or oxygen, resulting in partial oxidation at high temperatures and pressures. Other solid fuels such as biomass can also be gasified.

Rather than burning, most of the carbon-containing feedstock is chemically decomposed in the gasifier. Subsequent steps are similar to those in natural gas reforming—a shift reaction that produces a hydrogen-carbon monoxide syngas, followed by hydrogen separation and purification.

Coal gasification potential

Coal gasification is also a potential route toward cleaner power generation, and the lowest-cost coal-based hydrogen plants of the future are likely to be ones that co-produce hydrogen and electric power, said NRC.

Gasification will increasingly be integrated with turbines for lower emissions and greater efficiency (Fig. 5–6). In integrated gasification combined-cycle (IGCC) systems, the syngas is cleaned and burned as fuel in a combustion turbine-generator. The turbine's exhaust heat is used in a boiler to create steam that powers a second turbine-generator.

Several gasification technologies have been developed and are in commercial use worldwide, and four IGCC plants are operating (two in Europe and two in the United States). ChevronTexaco alone has 65 gasification facilities in operation and produces 5 billion scf/d of syngas. The company sold its gasification technology business to General Electric in 2004. American Electric Power Co. plans to build a 1,000-megawatt IGCC plant by 2010. The facility would be four times larger than existing plants in the United States.

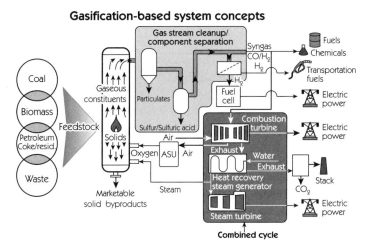

Fig. 5–6. Gasification of coal and other hydrogen feedstocks. Coal and other solid fuels can be gasified to produce hydrogen, electric power, and valuable by-products. Artwork courtesy of U.S. DOE.

Even with advanced technologies, however, hydrogen production from coal gasification would release about twice as much carbon as natural gas reforming does, said NRC. Carbon capture and storage could reduce these emissions by about 85%.

Hydrogen production economics

The cost of adding carbon capture to a coal gasification plant is only about 7% to 8% of the cost of the hydrogen produced. However, carbon storage costs are relatively high due to the greater volumes involved, compared to natural gas reforming. Accounting for these factors results in production of hydrogen at a cost of $1.23/kg with today's processes and $0.95/kg with future technologies (Table 5–2).

TABLE 5–2. HYDROGEN COSTS FOR CENTRAL COAL GASIFICATION PLANT

Technology	Hydrogen cost, $/kg H_2						
	Without carbon capture			With carbon capture and storage			
	Plant	Carbon emissions[*]	Total	Plant	Storage[**]	Carbon emissions[*]	Total
Current	0.96	0.26	1.22	1.03	0.16	0.04	1.23
Future	0.71	0.23	0.94	0.77	0.15	0.03	0.95

[*] At carbon emissions costs of $50 per metric ton.
[**] At carbon storage costs of $37 per metric ton.
Source: NRC.

In addition to carbon sequestration, hydrogen production from coal would depend on advances similar to those required for steam reforming of natural gas—lower costs, better catalysts, advanced membrane separation technology, and integration of some of the process steps.

Biomass and Waste Feedstocks

Biomass and waste include a great variety of materials—inedible plants, undesirable species that create a nuisance, plants grown specifically as biomass crops (Fig. 5–7), leftovers from agricultural crops, organic industrial waste such as wood chips from sawmills, animal waste from farming operations, solid waste collected in urban areas, and gases generated from landfills and sewage treatment plants. Biomass can be considered a renewable energy source in the sense that it was created with solar energy and can be replaced by growing more plants.

FIG. 5–7. BIOMASS CROP OF WILLOW TREES. AT THIS BIOMASS RESEARCH FARM IN NEW YORK, THE TREES ARE HARVESTED EVERY THREE YEARS, PROVIDING A RENEWABLE SOURCE OF HYDROGEN AND/OR ELECTRIC POWER. PHOTO BY DANIEL PECK, COURTESY OF NREL.

Solid biomass can be gasified to make biogas (biological synthesis gas), which can be processed into hydrogen. Many forms of biomass such as animal waste are suitable for a biological digestion process called anaerobic fermentation (decomposition in an oxygen-free tank). Once the biogas has been generated, subsequent steps to produce hydrogen are similar to those used in natural gas reforming and coal gasification.

Gasification of biomass to produce electricity and heat is a well-proven process, with commercial plants in operation worldwide. Biomass can also be incinerated directly or used as a supplemental fuel in coal combustion and gasification (up to 15% of total heat input).

In some cases, instead of gasifying the biomass, the material is pyro-lyzed—a decomposition process that produces usable liquids (bio-oils). These liquids can yield not only hydrogen but also valuable fuels, such as biodiesel and ethanol, and chemicals such as phenolic resins, which are used to make adhesives and other products. Ethanol can be converted to hydrogen in large refineries, and scientists are working on small ethanol processors that would produce just enough hydrogen for home heating and fuel cell vehicles.

Carbon emissions

From an environmental standpoint, biomass's attraction is that the fuel is carbon-neutral. Plants absorb CO_2 from the air while growing and fix it in solid form; when the plants are burned or otherwise consumed, they cannot release more CO_2 than they absorbed in the first place. Also, the CO_2 emitted when biomass fuels are used could be recaptured by new plant growth as part of nature's carbon cycle. If carbon emissions are captured and stored when biomass fuels are used, net CO_2 emissions could be negative.

However, if the plants are not regrown, the CO_2 is never recaptured. Also, cultivation of biomass consumes fuel for planting, irrigation, fertilizing, harvesting, and transportation.

Hydrogen production economics

Producing hydrogen from biomass is thermodynamically inefficient and expensive. Only a minuscule amount of the solar energy in biomass (0.2% to 0.4%) is converted to hydrogen. Gasification plants dedicated to biomass are limited to medium sizes due to the diversity of biomass feedstocks, their localized production, and the high costs of gathering and transporting them, said NRC.

Biomass gasification in a mid-size plant using today's technologies produces hydrogen costing more than $7/kg. Research to improve conversion efficiencies, along with related R&D on producing hydrogen from syngas, could reduce the cost of biomass-derived hydrogen to less than $4/kg.

Also, dedicated biomass crops would compete for land area with agriculture, conservation, recreation, and other land uses. If future technologies increase biomass crop yields by 50%, and if overall processing efficiencies are boosted to 40%, biomass crops would still require about 40% of the agricultural cropland in the United States to fuel an all-hydrogen fleet of vehicles, said NRC. However, cofiring biomass with coal in gasification plants could help mitigate CO_2 emissions and, with carbon sequestration, could result in theoretically negative emissions, or a reduction in atmospheric CO_2.

Electrolysis of Water

During electrolysis, an electrical current passes through water and splits it into hydrogen and oxygen (see chapter 3). Traditionally, two kinds of electrolyzers have been operated in industry, both using liquid electrolytes—the tank type (unipolar), with electrodes suspended in a tank of electrolyte, and the filter-press type (bipolar), which uses two separate cells and is better suited to high temperatures and pressures. The efficiency of these systems has improved from the 70% to 75% range in the 1970s to 80% to 90% today (based on the higher heating value of hydrogen), but essentially the concept hasn't changed for more than 50 years.[7]

Recently, however, electrolysis using PEMs has captured attention. The technology, originally developed in the 1970s, is the reverse of that used in PEM fuel cells. A solid polymer electrolyte (thin plastic sheet) becomes conductive when soaked in water. PEM electrolyzers reach efficiencies similar to conventional types, but they generate hydrogen at very high purities and at high pressures, reducing subsequent compression costs. Also, PEM units can handle abrupt variations in electrical input, which can disrupt conventional electrolyzers.

Steam electrolysis is a variant process where some of the energy needed to split the water is added as heat instead of electricity, which improves efficiency. Using concentrated solar energy to reach the very high temperatures at which steam decomposes is discussed later in this chapter (see *Solar water-splitting*). The problem here is to prevent the hydrogen and oxygen from recombining.

An intriguing concept being developed is to reverse the operation of a solid oxide fuel cell and make a hybrid electrolyzer/fuel cell system that could operate in either mode. A system being tested uses a small stack of high-temperature solid oxide cells that can produce either electricity from fuel or hydrogen from water. The concept is to burn a small amount of natural gas inside the cell to promote oxygen ion migration and reduce the electrical energy required, possibly reaching higher net electrochemical efficiencies.

Available systems

As with the reforming industry, electrolyzer units (often called hydrogen generators) are available in a range of sizes to serve hydrogen fueling stations, and two types are being scaled down for home use. Many electrolyzer systems are being demonstrated with fuel cell vehicle fleets, including a Norsk Hydro unit at a Hamburg fueling station (Fig. 5–8).

Five manufacturers dominate the global electrolyzer market and produce systems with varying hydrogen production capacities suitable for particular fueling station sizes (Table 5–3).

In addition to these manufacturers, Mitsubishi plans to begin marketing electrolyzers in 2005 through a Canadian subsidiary.

Fig. 5–8. Electrolyzer at Hamburg hydrogen fueling station. This station, which gets its hydrogen from a Norsk Hydro unit, serves three fuel cell buses and other hydrogen vehicles. Photo ©Norsk Hydro.

Table 5–3. Available electrolyzers for hydrogen fueling station applications

Company	Hydrogen production rate		Cars served	Station size
	kg/d	kg/yr		
Norsk Hydro	22–1,046	7,875–381,864	39–1,909	Small neighborhood (5–50 cars) up to commercial fueling station (>500 cars)
Stuart Energy	6–194	2,364–70,863	12–354	Small neighborhood to commercial fueling station
Teledyne	6–91	2,205–33,069	11–165	Small neighborhood to small fueling station (150–500 cars)
Proton Energy	1–22	396–7,875	2–39	Home (1–5 cars) to small neighborhood
Avalence	0.9–11	315–3,939	1.6–20	Home to small neighborhood

Source: U.S. DOE[8]

Hydrogen production economics

An economic analysis of these products showed that electricity prices will continue to be a major component of the cost of electrolytic hydrogen regardless of technology improvements, although the smaller the fueling station, the more significant becomes the capital cost of the electrolyzer.[9]

Electricity represents 80% of hydrogen costs at a full-scale commercial station and 62% at a small fueling station. In the case of a neighborhood station, capital cost rises to 43% of the total, but electricity still accounts for 35% of the hydrogen's cost. In general, at today's electrolyzer efficiencies, electricity prices would have to be below 4¢–5.5¢/kWh to produce hydrogen costing less than $3/kg.

Overall, the costs of hydrogen from PEM systems and from traditional liquid-electrolyte systems are roughly comparable, said NRC, and electrolyzers could play a role in the early stages of a transition toward hydrogen energy systems. The units scale down well, since their efficiency is independent of the size of the cells. Electrolyzers are compact enough to install at existing fueling stations, and they exploit the existing infrastructure for delivering water and electricity. They also might serve as off-peak hydrogen generators in micropower applications.

Reducing the capital cost of PEM electrolyzers depends on the success of PEM fuel cell R&D, which would translate to cost reductions in both types of systems. Technology advances could increase electrolyzer efficiency from 63.5% to 75% (based on hydrogen's lower heating value). These improvements might include minimizing parasitic energy consumption and reducing the current density required.

With R&D success, electrolyzer costs could fall by a factor of eight, said NRC, from $1,000/kW in the near term to $125/kW by 2020. At these prices, electrolyzers could produce hydrogen costing $4/kg using grid electricity, making electrolysis attractive during a transition through 2030.

However, using fossil-fueled grid power for electrolysis would wipe out much of the environmental benefit of using hydrogen in light-duty fuel cell vehicles. No net carbon reduction would be achieved, compared to a U.S. fleet of gasoline-electric hybrid vehicles, if onsite electrolyzers produced the hydrogen using grid power. Similarly, onsite electrolysis using the European power mix results in some of the highest greenhouse gas emissions and energy consumption of any hydrogen pathway.[10]

Integrating electrolysis with renewables

Electrolysis might be considered the only practical link between renewable energy and usable fuels. Energy from the sun, wind, and other renewable sources has to be converted into chemical energy for automotive and stationary use. Generating electricity from renewable sources and using it to make hydrogen via water-splitting offers a way to store this intermittent power and use it on demand in hydrogen fuel cells.

However, research is needed to reduce the costs and improve the performance of systems that integrate electrolysis with renewable power sources (see *Hydrogen From Solar Power* and *Wind-Powered Hydrogen Production* in this chapter). One example is to optimize power conversion equipment and other system components.

If onsite electrolyzer capital costs are reduced to $300/kW and efficiency increased to 73%, hydrogen could be produced for $2.50/kg from an integrated renewable power/electrolysis system producing 250 kg/d of hydrogen at 5,000 psi (340 bar), said NRC. Research is underway to integrate Avalence's Hydrofiller technology, which can produce hydrogen at 10,000 psi (1,450 bar), with solar photovoltaic (PV) arrays, without using power conditioning equipment.

Another alternative might be to use much larger electrolyzers, which already have lower capital costs per kilowatt, power them with the entire output from a wind farm, and let them sit idle when the wind isn't blowing rather than using grid electricity.

Nuclear Power

Some people think that nuclear power's heyday has faded into oblivion, but instead this primary energy source is staging a comeback worldwide, or trying to. Nuclear electricity is the only existing large-scale option for carbon-free electrolysis of water into hydrogen, yet this advantage is tempered by the industry's record of accidents, the toxic waste storage dilemma, a history of over-budget construction, and its close association with weapons of mass destruction.

Scientists are working on ways to make our next generation of nuclear power plants safer, cheaper, and more environmentally acceptable. Although the vision of nuclear's newborn advocates might be too optimistic, a vociferous debate will continue on nuclear's position in the world's energy mix, and R&D will probably lead to construction of more advanced technology demonstration plants in industrial nations.

Industry status

In Asia, nuclear power is enjoying a growth trend, particularly in China, India, and Japan. Of 27 new plants under construction worldwide, 18 are in Asia. Globally, 442 existing plants provide about 16% of our electricity supplies while emitting only a few grams of carbon per kilowatt-hour, on a par with renewable sources. Japan operates one of the world's most advanced nuclear power complexes.

European countries differ in their nuclear energy policies. Finland is planning a new plant, and France—often considered a model of modern nuclear power—might soon begin replacing its aging facilities. However, nuclear power is being phased out in the Netherlands, Germany, Belgium, and Sweden.

The U.S. government is funding two international research programs with multiple partners: Generation IV, which aims to develop new fission reactor designs, and ITER, which focuses on nuclear fusion technology. The GenIV project might include construction of a nuclear facility in Idaho that produces both electricity and hydrogen.

Hydrogen production economics

Opponents point out that regardless of nuclear power's dangers, plants would have to be built and operated much more economically to compete in the future. Vijay Vaitheeswaran analyzed the nuclear industry in *Power to the People* and concluded, "Barring some extraordinary technological break-through, the industry that once boasted it would be too cheap to meter is likely to be remembered as too costly to matter."[11]

A Shell study suggested that hydrogen production from nuclear energy is not an efficient process, either by electrolysis or by direct thermochemical water-splitting (using heat and chemicals). Shell favors fossil fuel feedstocks and carbon sequestration.

Technologies being developed

Today's water-cooled nuclear reactors generate electricity that can be used to produce hydrogen via electrolysis, but they do not operate at high enough temperatures to produce hydrogen directly via thermochemical water-splitting. Gas-cooled reactors are being designed that could make hydrogen more efficiently by operating at higher temperatures (700° to 1,000°C). These plants might electrolyze steam or thermochemically split water. As another competing option, the heat could be used to reform natural gas.

One drawback of high-temperature operation is the need for cost-effective, durable materials. Ultrahigh-temperature processes are also being studied (1,000° to 3,000°C) for direct thermochemical water-splitting. Several processes are summarized in Table 5–4.

TABLE 5–4. NUCLEAR HYDROGEN PRODUCTION OPTIONS

	Electrolysis		Thermochemistry	
	Water	Steam	Methane reforming	Water-splitting
Temperature, °C	>0	>300 (LWR) >600 (S-AGR)	>700	>850 (S-I) >600 (Cu-Cl)
Process efficiency, %	75–80	85–90	70-80	>45
Efficiency coupled with LWR, %	27	30	--	--
Efficiency coupled with HTGR, AHTR, or S-AGR, %	<40	40–60	>70	40–60

LWR=light-water reactor
S-AGR=supercritical advanced gas reactor
S-I=sulfur-iodine cycle
Cu-Cl=copper-chlorine cycle
HTGR=high-temperature gas reactor
AHTR=advanced high-temperature reactor

Source: NRC

Nuclear fusion technology is regarded as a long shot, but is not being neglected by the research community. Fusion would produce no radioactive fuel waste or weapons-grade materials. And fusion's fuel source—seawater—is plentiful. The reactor would use the deuterium hydrogen isotope extracted from seawater and the tritium isotope, which is made from lithium, also a seawater component. The technical challenge is controlling the thermo-nuclear reaction.

Hydrogen From Solar Power

Energy from the sun consists of both light and heat. PV systems convert the sun's light energy directly into an electric current. In PV systems, incoming sunlight is absorbed by semiconducting materials, allowing electrons from their atoms to flow through an external circuit and generate electricity. The greater the intensity of the light, the more power is generated.

The basic building block of a PV system is the solar cell containing the semiconducting material, most commonly crystalline silicon, although many new materials are being developed. An individual solar cell typically produces one to two Watts of electricity. The cells are connected into modules (usually containing about 40 cells) and enclosed in protective casings. PV modules can be wired together into large arrays that produce hundreds of kilowatts. Today's PV materials can also be fashioned into panels, canopies, roofing shingles, and other integrated building elements.

Solar energy collected by PV panels and canopies can be used to power electrolyzers and supply pollution-free hydrogen to fueling stations and other applications. This technology is fairly well-proven, but not yet economical due primarily to the high cost of solar modules.

One of the first examples of PV-powered hydrogen was Hysolar, a co-operative project between Germany and Saudi Arabia during the 1980s and 1990s. In a solar energy complex near Riyadh, low-pressure water electrolyzers were powered by various types of PV panels. Hysolar reportedly reached an overall electrolysis efficiency of 69%.[13]

More recently, a Stuart electrolyzer at the Barcelona hydrogen fueling station is being powered partially by onsite PV panels, as part of Europe's CUTE fuel cell bus demonstration (see chapter 4). Stuart is working with partner Solar Integrated Technologies, a PV canopy manufacturer, to demonstrate other small PV-powered hydrogen fueling stations.

Honda opened a solar-powered station in California in 2004, and Illinois is planning a solar and wind-powered station at a regional airport near Chicago as part of its Hydrogen Highway program.

Australian researchers reported that cheap, light-sensitive ceramics can harvest sunlight and split water to produce hydrogen. The titanium oxide photoelectrodes have the right semiconducting properties and high resistance to corrosion.

Hydrogen production economics

Today's solar PV modules cost much less than they used to, but are still expensive at $3 to $6/Watt (at peak output). To compete with conventional power generation alone, not including hydrogen production, PV modules would have to cost less than $1/Watt, said NRC. Research on new PV cell manufacturing processes and on organic polymer-based solar cells holds promise for mass production at low cost.

At $1/Watt, PV systems could generate electricity costing 9.8¢/kWh and produce hydrogen via electrolysis at a total cost of $6.18/kg. This assumes sunlight availability of 20% and capital costs of $1.54/kg for electrolysis, hydrogen storage, and fuel dispensing.

Hydrogen at this cost would be attractive for small onsite applications, but cannot come close to competing with other options for large-scale production. Also, when solar energy is insufficient for the application, the system would have to use backup grid power, negating some of its carbon emission-reduction benefits.

Solar water-splitting

High-temperature thermochemical water-splitting can be powered by solar energy rather than a nuclear reactor. The idea is to concentrate the sun's heat intensely, using an array of mirrors on the ground aimed at a tower. Heated to thousands of degrees, steam dissociates into hydrogen and oxygen. In addition to materials research, solar energy collectors would have to be developed that use less land area and are much less expensive.

Scientists in Arizona have produced hydrogen from water at only 850°C using a proprietary solar thermal chemical process. Sunlight is concentrated by mirrors using a new collection technology and solar tracking software developed by SHEC-Labs (Solar Hydrogen Energy Corp.).

Photolytic processes

Photolysis simply means that light is the active force driving electrolysis. Photolytic processes for producing hydrogen include photoelectrochemical and photobiological techniques.

In photoelectrochemistry, semiconductors immersed in liquid electrolyte receive sunlight, resulting in a photocurrent within the cell that splits the water molecule. In addition to absorbing the solar energy, the semiconductor functions as an electrode. Sunlight-induced corrosion is a problem, and more durable materials would be needed for practical systems.

Another type of photoelectrochemical process uses a soluble metal complex as a catalyst. When it dissolves, the complex absorbs solar energy and produces an electrical charge that drives the water-splitting reaction. This method mimics nature's process of photosynthesis.

A milestone in photoelectrochemical hydrogen generation was reported in 1972 by two Japanese scientists who produced small amounts of hydrogen by shining light on semiconductor electrodes, without additional voltage. Then the U.S. National Renewable Energy Laboratory (NREL) achieved a conversion efficiency of 12.4% in 1998 by splitting water directly with sunlight in a single step, using a combination of photovoltaic and photoelectrochemical cells.[14]

Photobiological processes

Researchers are looking at ways to harness photosynthesis for hydrogen production via photobiological (or biophotolytic) processes using algae and bacteria. Under specific conditions, the pigments in certain types of algae absorb solar energy. An enzyme in the algal cell acts as a catalyst to split the water molecule and release hydrogen.

The scientific challenge is that this enzyme is sensitive to oxygen, so work is focusing on genetic engineering of algae to produce oxygen-tolerant mutants (Fig. 5–9). Some bacteria are also capable of producing hydrogen, but unlike algae they require a substrate to grow on.

In experiments at the University of California-Berkeley, an algae culture grown under normal conditions was deprived of sulfur and oxygen, causing it to switch to an alternative metabolism that generates hydrogen. After several days of hydrogen production, the algae culture was returned to normal conditions for a few days, allowing it to store up more energy.

Fig. 5–9. Algae for biological hydrogen production. Scientists are attempting to produce hydrogen using genetically engineered algae and other microorganisms. Photo courtesy of University of California–Oakland.

This process, which could be repeated many times, might eventually provide a cost-effective and practical means to convert sunlight into hydrogen. Scientists are also attempting to increase the light conversion efficiency of algae.

Researchers at the Imperial College London reported in *Science* in 2004 that X-ray crystallography has revealed new details about the protein complex that drives photosynthesis, including the atomic structure of the catalytic center and the position of key amino acids.[15]

Wind-Powered Hydrogen Production

Electricity generated by wind turbines is already much cheaper than solar power and closer to becoming an economical way to produce pollution-free hydrogen in the future. In some areas, wind-generated electricity is already cost-competitive with new fossil-fueled plants. At good wind sites (Class 4 or better), electricity can be generated at costs of 4¢ to 7¢/kWh (without incentives), when the turbines are running 30% of full power on average.

Global wind capacity approaches 40,000 megawatts, mostly in Europe and the United States, with Germany and Spain the top two countries. Shell is the most prominent multinational involved in wind energy. The Wind-Energy division of Shell Renewables has a stake in 662 megawatts of wind capacity and plans to assemble a portfolio of 1,600 megawatts by 2006.

Small-scale wind-powered hydrogen production is being demonstrated at vehicle fueling stations and in experimental settings like the UK's Hydrogen and Renewables Integration (HARI) project, where two wind turbines contribute 50 kW toward electrolytic hydrogen production, along with 13 kW of input from solar PV arrays. In the United States, two 1.5-megawatt wind turbines to be installed on a university research farm in Iowa will generate electricity for direct use on-peak and for hydrogen production off-peak.

Norway's wind-to-hydrogen project

The Norwegian island of Utsira began producing hydrogen from wind power in 2004, making the island self-sufficient in energy use. The system incorporates two wind turbines and a Norsk Hydro electrolyzer, along with hydrogen storage equipment and a fuel cell for generating power on demand (Fig. 5–10).

FIG. 5–10. NORWAY'S WIND POWER-HYDROGEN PROJECT. POWER FROM THESE TWO WIND TURBINES PRODUCES HYDROGEN ELECTROLYTICALLY, WHILE A FUEL CELL PROVIDES ELECTRICITY ON DEMAND, MAKING THE NORWEGIAN ISLAND OF UTSIRA SELF-SUFFICIENT IN ENERGY USE. PHOTO ©NORSK HYDRO.

The island has only three or four windless days each year, but winds can become too strong or too variable to provide a steady supply of electricity. Hydrogen storage combined with the fuel cell solves the problem. A single wind turbine provides plenty of power for the island's households. Electricity from the second turbine is exported to the mainland.

Spain also is developing a wind-to-hydrogen project at the Unversidad Publica de Navarra, with support from Corporacion Energia Hidroelectrica de Navarra. A 5-kW electrolyzer supplied by Stuart Energy is fed by a 10-kW electronic converter that supplies voltage and current similar to those produced on a wind farm under varying operating conditions.

Outlook

Wind energy is the most affordable renewable technology, the closest to widespread use, and the one with the best chance of producing substantial amounts of hydrogen during a transition period and in the future, said NRC. Technology improvements are expected to continue to drive down costs and enhance turbine performance. Potential barriers include siting issues and the intermittent nature of wind power generation.

Integrating wind turbines with electrolytic hydrogen systems offers potential for optimization. Hydrogen storage can be tailored to wind turbine design, and as described earlier, the capital cost of electrolyzers is expected to fall while their efficiency increases. This means that tomorrow's small, onsite wind-hydrogen systems would not require backup electricity from the grid. The electrolyzer could be sized large enough that sufficient low-cost hydrogen is produced from wind power alone.

Hydrogen production economics

With existing technology, small-scale wind power can produce hydrogen costing $6.64/kg, according to NRC (large wind farms were not considered in its analysis). This assumes the use of backup power to alleviate under-utilization of the electrolyzer. With anticipated technology improvements, wind-powered hydrogen production costs could be reduced to $2.86/kg without grid backup (Table 5–5).

TABLE 5–5. WIND-POWERED HYDROGEN COSTS

	Current technology		Future technology	
	With grid backup	Without grid backup	With grid backup	Without grid backup
Average electricity cost, ¢/kWh	6	6	4	4
Wind turbine capacity factor, %	30	30	40	40
Hydrogen cost, $/kg	6.64	10.69	3.38	2.86
Carbon emissions, kg C/kg H_2	3.35	0	2.48	0

Source: NRC

Other Renewables

In countries with abundant water, hydropower is the leading renewable source of electricity, which is generated cheaply and with little air pollution. The energy of the global water cycle is driven by the sun, which constantly evaporates huge amounts of water. However, damming rivers and streams affects the habitats of local plants, fish, and animals.

Hydropower generates roughly 20% of the world's electricity. Although many of the best hydropower sites have been exploited, China is building the planet's largest dam, called the Three Gorges, and small-scale hydropower projects are under way worldwide, bringing electricity to countries that lack other energy resources. Countries with abundant hydropower, such as Canada and Brazil, could become leaders in implementing large-scale hydrogen production based on renewable electrolysis.

Geothermal energy is another sustainable energy resource, but has been less developed than hydropower. The heat generated by the earth's core is virtually unlimited, and its tappable sources include geysers, hot springs, and volcanic rock. Hot water that is pumped out and used for heating can be reinjected to maintain reservoir pressure.

Geothermal energy plants have operated for decades in Italy, New Zealand, and California, demonstrating the sustainable character of the resource. The electric production potential of U.S. geothermal sites is estimated at nearly 23,000 megawatts, according to the Geothermal Energy Association. A total of 23 states use geothermal resources.

Iceland enjoys an abundance of both hydropower and geothermal energy. From 1940 to 1975, Iceland shifted all of its space heating equipment from oil to geothermal heat. More recently, air pollution has prompted the island to develop the world's first hydrogen economy. Plans are to switch all of its cars, trucks, buses, and ships to hydrogen power by 2050 (see Chapters 2 and 4).

A forgotten stepchild of the 1970s energy revolution is ocean thermal energy conversion (OTEC). Our oceans are the world's largest solar energy collector and energy storage system. OTEC technology exploits the temperature difference between the warm surface of the sea and its colder depths. Hawaii and Japan have built OTEC plants producing less than 100 kilowatts, but research has languished.

A related concept is tidal energy, which recovers power from the ebbing and flowing of the oceans' tides. Canada, Norway, Scotland, and the Azores have operated tidal energy plants, which incorporate dams at the entrance to tidal basins. The UK's Department of Trade and Industry is funding $90 million (£50 million) in tidal energy research.

Some scientists are also seeking ways to generate power from the oceans' waves, which is estimated in the millions of megawatts. A 750-kW wave energy converter called Pelamis began providing power to the UK grid in 2004. In the United States, a 500-kW pilot wave-energy project is planned for Rhode Island.

Long-Term Hydrogen Supplies

The International Partnership for a Hydrogen Economy comprises 15 nations—Australia, Brazil, Canada, China, France, Germany, Iceland, India, Italy, Japan, Korea, Norway, Russia, the United Kingdom, and the United States, plus the European Commission—which are advancing the transition to a global hydrogen economy. An adequate supply of hydrogen over the long run will be a concern. At one end of the spectrum, Iceland hopes to export hydrogen to Europe, while Japan will likely need to import primary energy supplies for hydrogen production or perhaps liquefied hydrogen by tanker.

As the world's largest consumer of transportation fuels, America must address the hydrogen supply issue sooner than other countries. The U.S. DOE has estimated that about 40 million tons of hydrogen would be consumed by a fleet of 150 million vehicles (75% of America's light-duty fleet), assuming that the baseline hydrogen fuel economy of 27.5 mpg is improved by a factor of 2.2. Currently about 9 million tons of hydrogen are produced annually in the United States.

Clearly, during a transition period of several decades, a mix of feedstocks and energy sources would provide hydrogen for U.S. vehicles, gradually shifting from conventional fossil fuels toward renewable resources and advanced hydrogen production technologies. For purposes of comparison, however, DOE projected the amount of each domestic resource that would be required to produce the entire volume of hydrogen needed for vehicles alone (Table 5–6).

TABLE 5–6. HYDROGEN PRODUCTION FROM U.S. DOMESTIC RESOURCES

Technology/ Resource	Needed for Hydrogen	Current Availability	Current Consumption	Consumption Factor Increase	Construction/ Footprint Required
Reforming and/or Partial Oxidation					
Natural gas	95 million tpy	28 billion tons recoverable	475 million tpy	1.2	400 dedicated plants (100 million scf/d)
Biomass	400-800 million tpy	800 million tpy residue/ waste + 300 million tpy dedicated crops	200 million tpy (3 quads for heat, power, and electricity)	2-4	400-600 dedicated plants
Coal	310 million tpy	126 billion tons recoverable (bituminous)	1100 million tpy (all grades)	1.3	280 dedicated plants
Water Electrolysis					
Wind	555 GW_e	3,250 GW_e	4 GW_e	140	Available capacity of North Dakota ($\geq$Class 3)
Solar	740 GW_e	2,300 kWh/ m^2-yr (SW U.S.)	<1 GW_e	>740	3,750 sq miles (approx. area of White Sands Missile Range, NM)
Nuclear	216 GW_e	N/A	98 GW_e	3.2	200 dedicated plants (1–1.2 GW_e)
Thermochemical water-splitting					
Nuclear	300 GW_{th}	N/A	0	N/A	125 dedicated plants (2.4 GW_e)

Source: U.S. DOE *Hydrogen Posture Plan*

References

1. National Research Council, *The Hydrogen Economy: Opportunities, Costs, Barriers, and R&D Needs*. The National Academies Press, 2004 (394 pages). ISBN 0-309-09163-2, www.nap.edu.

2. Walters, Russell C., "Fuel Cell Tutorial," Iowa State University, http://erl.cce.iastate.edu.

3. National Research Council, *The Hydrogen Economy: Opportunities, Costs, Barriers, and R&D Needs*. The National Academies Press, 2004 (394 pages). ISBN 0-309-09163-2, www.nap.edu.

4. Lord Browne, "Climate Change," Institutional Investors Group, Bishopsgate, London, November 26, 2003, www.bp.com.

5. Moritis, Guntis, "CO_2 sequestration adds new dimension to oil, gas production," *Oil & Gas Journal*, March 3, 2003, pages 39–44.

6. Hoffman, Peter, *Tomorrow's Energy: Hydrogen, Fuel Cells, and the Prospects for a Cleaner Planet*. The MIT Press: Cambridge, MA, 2001. ISBN 0-262-08295-0, page 74.

7. Hoffmann, pages 61–62.

8. Ivy, Johanna, "Summary of Electrolytic Hydrogen Production: Milestone Report for the Department of Energy's Hydrogen, Fuel Cells, and Infrastructure Technologies Program," National Renewable Energy Laboratory Technical Report NREL/MP-560-35948, April 2004, page 3.

9. Ivy, page 1.

10. "GM Well-to-Wheel Analysis of Energy Use and Greenhouse Gas Emissions of Advanced Fuel/Vehicle Systems – A European Study," September 27, 2002, www.lbst.de/gmwtw.

11. Vaitheeswaran, Vijay V., *Power to the People: How the Coming Energy Revolution Will Transform an Industry, Change Our Lives, and Maybe Even Save the Planet*. Farrar, Straus and Giroux: New York, 2003. ISBN 0-374-23675-5, page 290.

12. Hoffmann, Peter, page 58.

13. Ibid., pages 75–78.

14. "Hydrogen Economy Breakthrough on the Horizon," February 13, 2004, www.solaraccess.com.

Storage and Safety

Hydrogen energy systems of the future will depend on safe, reliable ways to store hydrogen. Although hydrogen is difficult to store, electricity is even more so. This vital distinction gives hydrogen an edge in competing as a future energy delivery service.

The biggest challenge will be storing enough hydrogen onboard fuel cell vehicles to allow the same travel range as their gasoline or hybrid counterparts—at least 300 miles (500 km)—without an unreasonable weight penalty and without taking up too much valuable cargo space. The benchmark for weight is a full tank of gasoline (17–18 gallons), weighing about 110 pounds (240 kg). Most of today's fuel cell vehicles carry containers of compressed hydrogen gas, but some designs use tanks of supercold liquid hydrogen.

Hydrogen will also have to be stored at fueling stations and micropower plants, technology parks and microgrids, regional distribution hubs, and eventually at central hydrogen production plants. In these stationary applications, the weight of the hydrogen storage system might not be as important as in vehicles, but its size might be limited to available building space.

Natural gas or other feedstocks might be stored in addition to hydrogen at vehicle fueling stations or other applications with onsite reforming systems. In applications that rely on electrolysis to supply hydrogen on demand, some hydrogen would have to be stored to meet demand swings and system interruptions.

Like gasoline, stored hydrogen can pose safety risks if containers leak into enclosed structures such as parking garages or in spaces with limited ventilation such as tunnels. Safety can also become an issue during the vehicle refueling process, especially when consumers or other unskilled people are handling the equipment. In general, however, hydrogen is neither more nor less inherently dangerous than conventional fuels. Some of its properties make it safer, while others make it more dangerous. With proper handling and controls, hydrogen can be as safe as other fuels in common use, or even safer.

Hydrogen Storage Today

Our state-of-the-art hydrogen storage technologies are compression and liquefaction. Although hydrogen holds more energy by weight than any known fuel, it contains less energy per unit of volume than competing fuels (see chapter 3). Because hydrogen takes up so much space, it has to be compressed or liquefied to store useful quantities. These physical storage techniques increase hydrogen's energy density. However, they also consume energy, contributing to the cost of the hydrogen supply chain.

Compressed gas tanks

Most compressed hydrogen tanks operate at ambient temperatures (normal weather conditions) and store the gas at pressures of 5,000 to 10,000 psi (350 to 700 bar, or 35 to 70 MPa). Compression of hydrogen to this range requires roughly 10% of the gas's energy content.

Compressed hydrogen tanks (usually cylinders with rounded ends) are available from many manufacturers and have been certified by standards agencies worldwide. They are in use onboard fuel cell vehicles, in portable and onsite power generators, and in uninterruptible power supply systems, where fuel cells replace batteries (see chapter 4).

Conventional steel tanks are too heavy for hydrogen storage onboard vehicles, especially when made of premium steel to prevent metal fatigue and leakage. Manufacturers have introduced lightweight cylinders that use a variety of other materials such as polymer liners, multiple shells, and composite fiber wraps to minimize the amount of metal required.

Liquefied gas tanks

Hydrogen liquefies at supercold (cryogenic) temperatures (−253°C, or −423°F). In its liquid state, hydrogen takes up just 1/700th as much volume as the gaseous form—an important advantage where storage tank size is a concern. Liquid hydrogen has a higher energy density than the compressed form, so a tank containing the same amount of fuel is smaller.

For a given volume of fuel, liquid hydrogen would boost the driving range of a vehicle, allowing roughly two to three times the distance of a compressed gas tank, making the liquid form ideal for long road trips. Also, liquefied hydrogen tanks are not highly pressurized and can be filled up faster than compressed gas tanks.

However, liquefaction requires much more work than compression, consuming at least 30% to 35% of hydrogen's energy content, or 11 to 12 kWh of electricity per kilogram of hydrogen liquefied. Also, cryogenic tanks require insulation and other special materials.

The other enemy of liquid hydrogen storage is evaporation. Whenever the hydrogen isn't being used—for example, when a fuel cell vehicle is parked—the liquid naturally tends to return to its gaseous state, due to heat exchange between the cold tank and the warmer atmosphere, and the boiled-off gas must be vented. Evaporation also occurs while the vehicle is moving and using fuel, but not as quickly.

Lawrence Livermore National Laboratory estimates that fuel cell vehicles lose liquid hydrogen to evaporation when they are driven less than 15 miles a day on average (25 km). Typically, boil-off begins within three to four days of a tank fill-up. When the vehicle is not driven at all, the entire tank of liquid hydrogen would evaporate within three weeks.[1]

Nevertheless, liquid hydrogen has its proponents, and cryogenic on-board hydrogen storage is being tested in many vehicles. BMW uses liquid hydrogen in a hybrid vehicle powered by a hydrogen-gasoline engine. Another idea being studied is a dual tank system that would use ambient-temperature, high-pressure hydrogen gas for short trips and the cryogenic liquid form for longer distances. Researchers at Lawrence Livermore are also working on hybrid tanks that might contain three forms of hydrogen—compressed gas, cryogenic liquid, and cryogenic compressed gas.

Advanced Storage Technologies

Compression and liquefaction are physical methods of storing molecular hydrogen. Newer technologies incorporate atomic hydrogen into the structure of metals and other materials. Hydrogen atoms bound into hydrides and other compounds cannot spill, vent, or burn, making them safer than conventional physical storage. However, developers are struggling to overcome the weight penalties associated with storing hydrogen in materials.

Three mechanisms bind hydrogen into storage materials: absorption, adsorption, and chemical reaction. An example of absorption is metal hydrides, where metals absorb atomic hydrogen into their crystalline structure. In the second mechanism, adsorption, hydrogen is bonded onto highly porous materials with a lot of surface area.

In the chemical route, a reaction takes place between the hydrogen and the storage medium. Reversible chemical reactions release hydrogen under small changes in temperature or pressure and can be cycled many times without losing capacity. So-called irreversible reactions are not actually irreversible, but do require additional chemical reactions or greater changes in temperature or pressure to release the hydrogen.

Metal hydrides

Originally, metal hydrides were designed to control reactions and output in nuclear power plants. Their development for hydrogen storage was prompted by their safety advantage. In this concept, a pure metal or an alloy forms a stable hydride when it absorbs the hydrogen atom and gives off heat. The hydride releases the hydrogen when heat is applied to reverse the absorption process (Fig. 6–1). Heat given off by a fuel cell can be used to release the hydrogen.

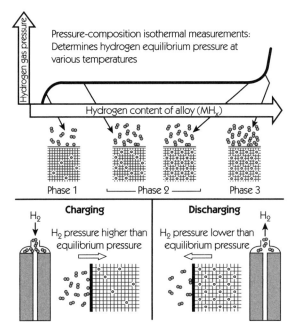

FIG. 6–1. METAL HYDRIDE STORAGE PROCESS. METAL ALLOYS ACT LIKE SPONGES TO ABSORB GASEOUS HYDROGEN INTO THE SOLID METAL, FORMING A HYDRIDE THAT RELEASES THE HYDROGEN WHEN HEATED. ARTWORK COURTESY OF TEXACO OVONIC.

When incorporated in a metal hydride, the hydrogen takes up no additional space, resulting in a compact, high-density storage medium. Compared with compressed gas tanks holding the same amount of hydrogen, metal hydrides are only one-third to one-fourth as large. Also, they operate at low pressures (less than 200 psi).

However, metal hydrides are heavy, and a tradeoff is involved in selecting materials. The metals with the highest storage capacity require high temperatures to release the hydrogen. This is not the best match for using the low-temperature exhaust heat from PEM fuel cells in vehicles, especially during rapid acceleration when the maximum amount of hydrogen needs to be released. But lower temperature hydrides hold less energy, so they have to be larger and heavier to carry as much hydrogen fuel. Researchers are working on new alloys that optimize the hydride's temperature and energy density characteristics.

Metal hydride storage technology is being demonstrated onboard fuel cell vehicles such as Toyota's Prius® and in stationary systems for uninterruptible power supplies (Fig. 6–2). The Prius's hydride technology was developed by Texaco Ovonic Hydrogen Systems, a joint venture between ChevronTexaco and Energy Conversion Devices, Inc. Dynetek Industries Ltd., a cylinder manufacturer, is working with Texaco Ovonic to develop a lightweight, high-pressure vessel incorporating metal hydride storage.

FIG. 6–2. RACK-MOUNTED HYDROGEN CYLINDERS. THIS REVERSIBLE METAL HYDRIDE STORAGE SYSTEM IS DESIGNED FOR REPLACING BATTERIES IN UNINTERRUPTIBLE POWER SUPPLY APPLICATIONS. PHOTO COURTESY OF TEXACO OVONIC.

Shell's partner in metal hydride research is HERA Hydrogen Storage Systems, which offers commercial low-temperature systems and is working on advanced technologies including reversible, high-temperature magnesium-based metal hydrides that can hold more hydrogen.

A leading Japanese producer of metal hydrides, The Japan Steel Works, Ltd. (JSW), is focusing on a titanium-chromium-vanadium alloy called BCC in addition to its commercially available alloys. JSW builds small metal hydride tanks and large, bulk storage systems.

Aluminum compounds called alanates are considered to be the most promising of the complex metal hydrides, and several research projects are under way to study their storage mechanisms.

Chemical hydrides

A chemical slurry or solution can store hydrogen as a hydride, releasing it through a reaction with water, which gives off heat. These systems require heat management and are irreversible in that the storage medium must be regenerated before it can be recharged with more hydrogen.

The most advanced chemical storage material is sodium borohydride—essentially, the familiar laundry detergent Borax (sodium borate) combined with hydrogen. The chemical creates a nontoxic, nonflammable solution with water and produces hydrogen when exposed to a catalyst. When the catalyst and solution are separated, the system stops generating hydrogen. The spent solution is recyclable.

Sodium borohydride storage is being developed by Millennium Cell in a system called Hydrogen on Demand®, which uses a proprietary catalyst (Fig. 6–3). The technology was demonstrated successfully in a Daimler-Chrysler fuel cell concept vehicle called the Natrium minivan, a Town & Country model. (Natrium is the Latin name for sodium.) The vehicle won the *Popular Science* "Best of What's New" award in 2002.

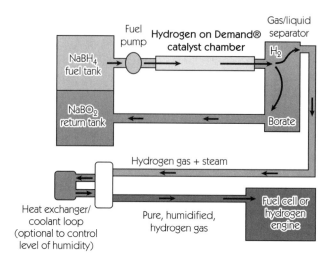

FIG. 6–3. BOROHYDRIDE STORAGE PROCESS. THE HYDROGEN ON DEMAND®
SYSTEM STORES HYDROGEN CHEMICALLY IN A SODIUM BOROHYDRIDE SOLUTION,
WHICH RELEASES THE GAS WHEN PUMPED THROUGH A CATALYST CHAMBER.
ARTWORK COURTESY OF MILLENNIUM CELL.

PSA Peugeot Citroën agreed in 2004 to explore Millennium's borohydride technology in fuel cell vehicles with support from French utility Gaz
de France. The Hydrogen on Demand system can be scaled to portable and
standby power applications.

HERA is also researching irreversible chemical hydrides, which would
provide a one-time supply of hydrogen in large capacities. The hydrogen
would be released by reaction with water, ammonia, or alcohol.

In another type of chemical hydride, a thick slurry uses an inert, stabi-
lizing liquid to protect the hydride from contact with moisture and make it
pumpable. When mixed with water, the slurry releases high-purity hydrogen.
However, the spent hydride has to be recovered and regenerated at a central
plant. Researchers are seeking to identify safe, stable, pumpable slurries and
better reactor designs for regenerating the spent slurry.

Carbon materials

Scientists are excited about the promise of carbon materials that can hold hydrogen in minuscule structures called nanotubes. (Technically, the nano prefix means one-billionth, but can refer to anything extremely small, at the atomic level.) Theoretically, these forms of carbon could store and release huge quantities of hydrogen, if practical structures and methods can be developed. As Vijay Vaitheeswaran comments in *Power to the People*, "If nanotube storage is perfected, the entire hydrogen economy would leap forward by decades."[2]

In nanotubes, carbon atoms are linked together in hexagons like chicken wire and rolled into single-layer tubes. Richard Smalley calls the structures "bucky tubes" because they're an elongated version of "bucky balls," which were named after geodesic dome creator Buckminster Fuller. Smalley shared a Nobel Prize in 1996 for discovering bucky balls (fullerenes).[3]

Carbon nanotubes are known to attract molecules of hydrogen, but researchers haven't figured out how to control its uptake and release. Also, they have no clue yet how to mass-produce the tubes (Fig. 6–4). In the laboratory, nanotubes are made by blasting carbon with a laser beam, which creates batches of material that collect in cobweb-like strings.

FIG. 6–4. CARBON NANOTUBE RESEARCH. SCIENTISTS ARE TRYING TO FABRICATE BUNDLES OF CARBON NANOTUBES THAT COULD STORE HYDROGEN IN LIGHTWEIGHT SYSTEMS FOR FUEL CELL VEHICLES. PHOTO BY JIM YOST, COURTESY OF NREL.

The huge surface area afforded by the carbon atoms offers immense potential for hydrogen storage, and the carbon atom is lightweight. The idea is that carbon nanotubes might draw in hydrogen just as water is sucked into a straw. Tube bundles might serve as light hydrogen sponges ideal for fuel cell vehicles. Also, nanotubes are extremely stiff and strong, and plenty of carbon is available to manufacture them.

Barriers and Challenges

The most obvious general obstacles to overcome in storing hydrogen are the weight and size of the container and its contents, especially in automotive applications. In addition, the costs of materials and manufacturing need to be reduced, and public acceptance will require proven safety for any kind of hydrogen storage system to enter the market.

So far, no compressed gas or liquid hydrogen technology is capable of storing enough hydrogen to meet the automotive industry's requirements for mass market success. For a travel range of 300 miles (500 km), the containers and fuel needed for a light-duty vehicle would be either too large or too heavy or both.

The time required to refuel the tank and the complexity of the process are also issues that need to be addressed for consumer acceptance. The situation is different for urban transit buses, which do not need to travel as far. Typically they carry hydrogen storage tanks on their roof and are refueled by skilled workers at a central location after their shift of duty.

Technological breakthroughs would be needed to lower the amount of energy required for liquefying hydrogen. This would reduce not only storage costs, but also the cost of delivering hydrogen by truck, rail, or ship (see chapter 7). In reversible hydride systems, heat management during the hydrogen storage and release processes needs to be studied, as well as weight penalties.

The principal shortcomings of chemical storage are the high cost of making the chemicals and the need to demonstrate technologies for recycling or disposing of waste products, said the U.S. NRC in its report, *The Hydrogen Economy: Opportunities, Costs, Barriers, and R&D Needs.*[4]

Nonetheless, NRC thinks that chemical storage in borohydrides and other compounds is an intriguing, "game-changing" possibility. In fact, the report recommended that the U.S. DOE stop funding research on compressed gas tanks and liquefied hydrogen storage for use onboard vehicles. Instead, research should continue to encourage novel concepts, because success in overcoming the major stumbling block of onboard storage is critical to the future of hydrogen use in fuel cell vehicles.

Most of the storage technologies being studied today apply only to relatively small amounts of hydrogen—enough to serve a large vehicle fueling station or maybe a small regional distribution hub.

In the future, if hydrogen applications become pervasive and much greater supplies are needed, central hydrogen production plants would have to store much larger volumes—up to five days' worth of output—to accommodate fluctuating demand and plant downtime. Possibilities include underground salt caverns, depleted gas fields, and empty mines. Huge quantities of natural gas are already stored in depleted fields, but research would be needed on their suitability for storing hydrogen.

Government/Industry Funding

Hydrogen storage research in the United States has been divvied into three Centers of Excellence, which are competing in what DOE calls the grand challenge. Over a five-year period, DOE is providing $150 million in funding to the centers, plus private cost-sharing estimated at $20 million. The groups are led by national laboratories, and each includes at least ten partners from industry and academia (Table 6–1).

TABLE 6–1. HYDROGEN STORAGE GRAND CHALLENGE

Centers of Excellence		
Chemical Hydrogen Center LANL/PPNL	Metal Hydride Center SNL	Carbon Center NREL
Los Alamos National Laboratory (Los Alamos, NM)	Sandia National Laboratories (Livermore, CA)	National Renewable Energy Laboratory (Golden, CO)
Pacific Northwest National Laboratory (Richland, WA)	Stanford University (Stanford, CA)	California Institute of Technology (Pasadena, CA)
University of Pennsylvania (Philadelphia, PA)	General Electric (Niskayuna, NY)	Duke University (Durham, NC)
UCLA (Los Angeles, CA)	University of Hawaii (Honolulu, HI)	Penn State University (University Park, PA)
University of California-Davis (CA)	California Institute of Technology (Pasadena, CA)	Rice University (Houston, TX)
Penn State University (University Park, PA)	Jet Propulsion Laboratory (Pasadena, CA)	University of Michigan (Ann Arbor, MI)
University of Washington (Seattle, WA)	HRL Laboratories (Malibu, CA)	University of North Carolina (Chapel Hill, NC)
University of Alabama (Tuscaloosa, AL)	University of Illinois (Champaign, IL)	University of Pennsylvania (Philadelphia, PA)
Rohm and Haas (Philadelphia, PA)	Univ. of Pittsburgh/Carnegie Mellon Univ. (Pittsburgh, PA)	Oak Ridge National Laboratory (Oak Ridge, TN)
Millennium Cell (Eatontown, NJ)	NIST (Gaithersburg, MD)	Lawrence Livermore National Laboratory (Livermore, CA)
Intematix (Moraga, CA)	University of Nevada-Reno (Reno, NV)	NIST (Gaithersburg, MD)
US Borax (Boron, CA)	Oak Ridge National Laboratory (Oak Ridge, TN)	Air Products (Allentown, PA)
	University of Utah (Salt Lake City, UT)	
	Intematix Corporation (Moraga, CA)	
	Brookhaven National Laboratory (Brookhaven, NY)	

Individual Projects		
Prime	**Partners**	**Research Area**
TIAX LLC (Cambridge, MA)	Gas Technology Institute (IL) Yale University (CT) University of Oklahoma (OK)	Lifecycle and cost analysis
University of Missouri (St. Louis, MO)	Pacific Northwest National Laboratory (WA)	New materials
University of Connecticut (Storrs, CT)	Pacific Northwest National Laboratory (WA)	New materials
Michigan Technological University (Houghton, MI)	None	Chemical hydrides
Gas Technology Institute (two projects) (Chicago, IL)	Superior Graphite Co. (IL) NEXGEN Fueling (MN)	Carbon Off-board storage
Alfred University (Alfred, NY)	Savannah River Technology Center (SC) Mo-Sci Corporation (MO) CERALINK (NY)	New processes
Carnegie Institute of Washington (Washington, DC)	None	New materials
Research Triangle Institute (Research Triangle Park, NC)	State Scientific Research Institute (Moscow, Russia) ATK/Thiokol Propulsion (UT)	Chemical hydrides
State University of New York (Syracuse, NY)	PoroGen, LLC (MA)	Carbon
TOFTEC, Inc. (Gainesville, FL)	University of Florida, Gainesville (FL)	New processes
University of Michigan (Ann Arbor, MI)	Northwestern University (IL) Los Alamos National Laboratory (NM)	New materials
University of Pennsylvania (Philadelphia, PA)	Drexel University (PA) NIST (MD)	Carbon
University of California– Berkeley (CA)	Lawrence Berkeley National Laboratory (CA)	New materials
University of California– Santa Barbara (CA)	Los Alamos National Laboratory (NM)	New materials

Europe's HyNet group of business leaders is working on a European Hydrogen Energy Roadmap that includes development of storage technologies. The group provides input to the European Union's research programs. The International Partnership for a Hydrogen Economy also is working on storage research.

Hydrogen Safety

Hydrogen has been produced in commercial quantities since World War II and has a long history of safe use in the chemicals and refining industries and in aerospace applications. During these decades of handling hydrogen, these sectors, along with industrial gas suppliers, have established guidelines for safe operating practices. Ongoing hydrogen vehicle and fueling station demonstrations worldwide are adding to our practical experience (see chapter 4 and Appendices A–C).

Comparison with other fuels

Hydrogen can be as safe as other fuels that we use today, or even safer, with proper handling and controls. All fuels pose some degree of danger, and the safe use of any fuel focuses on preventing situations where the three combustion factors are present—an ignition source (spark or heat), an oxidant (air), and the fuel itself.

In general, hydrogen is neither more nor less inherently dangerous than common fuels. Some of hydrogen's properties make it safer to handle and use—for example, buoyancy. Hydrogen is so much lighter than air, or more buoyant, that it rises rapidly when released, creating turbulent diffusion and reducing its concentration below the lower flammability limit. This means that it is virtually impossible for hydrogen to explode in an open area. Also, its explosive power is much weaker than that of gasoline vapor.[5]

On the other hand, hydrogen can ignite more easily than gasoline or natural gas, and it has a wide range of flammable concentrations in air. This means that adequate ventilation and leak detection are critical in designing safe hydrogen systems. Table 6–2 summarizes hydrogen's properties compared to those of natural gas, propane, and gasoline.

TABLE 6–2. PROPERTIES OF HYDROGEN AND OTHER FUELS

	Hydrogen	Natural gas	Propane	Gasoline vapor
Density relative to air	0.07	0.55	1.52	4.0
Molecular weight	2	16	44	107
Density, kg/m^3	0.084	0.651	1.87	4.4
Diffusion coefficient, cm^2/s	0.61	0.16	0.12	0.05
Explosive energy, MJ/m^3	9	32	93	407
Flammability range, vol%	4 to 75	5 to 15	2 to 10	1 to 8
Detonation range, vol%	18 to 59	6 to 14	3 to 7	1 to 3
Minimum ignition energy, mJ	0.02	0.29	0.26	0.24
Autoignition temp. in air, °C	565-581	539	454-509	257
Flame speed, cm/s	346	43	47	42

Source: U.S. DOE; NRC

Safe is a relative term, and it does not mean incident-free. For purposes of comparison, we can look at the safety record of other fuels. Hundreds of deaths and thousands of injuries have been attributed to fuel transmission, distribution, and use in the United States, even though these fuels are considered safe and do have a good track record. Breakthrough Technologies Institute summarized fuel-related incidents (Table 6–3) and noted that other costs are not quantified, such as environmental damage caused by oil spills.

TABLE 6–3. SAFETY RECORD OF FUELS AND ELECTRICITY

Fuel/ energy	Fatalities	Injuries	Property damage	Period
Natural gas	360	1,596	$668 million	1986–2003
Gasoline	39	324	$875 million	1986–2003
Propane	41	290	$31 million	1986–2003
Electricity	684	1,184	$669 million	1998 (electrical fires) and 2000 (electrocution)

Source: Breakthrough Technologies Institute[6]

One important difference between hydrogen and conventional fuels is that a hydrogen flame produces very little radiant energy—one-tenth that of a hydrocarbon fire—and no sooty particles, smoke, or poisonous emissions. In contrast, when gasoline burns, radiant heat, smoke, and hot soot reach far beyond the flame, and the by-products of combustion can be deadly.

Commercially available compressed gas tanks are rigorously tested in automotive situations such as crashes, fires, and overfilling (burst test), and are even used as target practice to see what happens when they're hit by armor-piercing bullets. Other tests include exposure to on-road vibration, extreme temperatures, and corrosive fluids, as well as thousands of fill-up cycles.

In one test, hydrogen was leaked from a vehicle tank at a rate of 3,000 cubic feet per minute and set on fire. During the course of the burn, sensors inside the vehicle measured temperature increases of only a few degrees, because the flame rose straight up from the tank and radiated very little heat.[7] In another test, a fuel tank fire caused no damage to a hydrogen-powered car, but destroyed a conventional gasoline car.[8]

Leak detection

Safety research is focusing on development of sensors for detecting hydrogen leaks, which do pose a safety concern for hydrogen and fuel cell systems. Leak sensors must be sensitive enough to provide a safe and reliable alarm system, as well as rugged, easily manufactured, and reasonably priced. Dependable sensors will be critical in automotive applications, where fuel cells operate in an enclosed space.

The safety of parking hydrogen-fueled vehicles in garages was modeled in 2004 by Parsons Brinckerhoff for the California Fuel Cell Partnership. The study addressed residential two-car garages, commercial multi-story parking lots (above- and below-ground), and commercial maintenance and repair service stations.[9]

According to the model results, leaking hydrogen in flammable concentrations would not extend beyond two feet immediately surrounding the vehicle. Typical airflow from natural or mechanical ventilation would dilute the hydrogen. In cases of low ventilation, hydrogen sensors in the vehicle's wheel wells would detect and control the leak with emergency shut-off valves. Integrating these sensors and valves in hydrogen vehicles will be one of the most important preventive measures to ensure safety, said the report.

Scientists are also seeking to develop chemical additives that would give hydrogen, normally an odorless gas, a smell that's detectable by humans, similar to the way natural gas is odorized today. Existing odorants aren't practical because they would ruin the catalysts used in hydrogen fuel cells. However, even if odorized, hydrogen would rise so quickly that it would escape detection at ground level.

Safety practices, codes, and standards

The public is already accustomed to the idea of common-sense safety in gasoline service stations, and precautions around hydrogen would be much the same—no smoking, no refueling with the engine running, and no use of cellphones around the fuel. The possibility of static electricity igniting hydrogen would have to be prevented by built-in grounding systems.

In the industrial gas business, Air Products and Chemicals, the global leader in supplying merchant hydrogen (gas for sale to customers), has published Safetygrams on hydrogen (see Appendixes F and G). The company also offers safety training programs and services to assist engineers, technicians, and other people involved in hydrogen supply and distribution, including fuel cell applications.

Ultimately, hydrogen will be as safe as the practices, codes, and standards that apply to its use. Safe operating practices have proved to minimize the known hazards associated with handling hydrogen, such as fire and explosion. Some of these hazards are addressed through operator training and proper system design; for example:

- Purging hydrogen systems with an inert gas such as nitrogen is required to avoid the formation of flammable hydrogen-oxygen mixtures.

- Adequate ventilation can minimize or eliminate the formation of combustible hydrogen-oxygen mixtures.

- Because hydrogen burns with an almost invisible blue flame, special flame detectors are required.

Many regulations, guidelines, and codes and standards have already been established for existing uses of hydrogen, and the systems and organizations are in place to establish new codes and standards for fuel cells and hydrogen fueling stations. Because these systems are complex and will be used in a wide range of applications, standards development organizations worldwide are cooperating to prepare for the commercialization of hydrogen and fuel cell technologies.

The International Code Council in 2004 approved new provisions in its International Fire Code that apply to hydrogen storage at fueling stations. Among these were two new options—one for storing compressed hydrogen in tanks on top of the station's overhead canopy, and the other for storing liquid hydrogen in insulated underground tanks. These new codes also allow stations to treat metal hydride storage systems the same way as gaseous systems.

The industry maintains a website that tracks global codes and standards that apply to hydrogen and fuel cells (www.fuelcellstandards.com). The information includes files for standards by application—stationary, transportation, portable/micro, and hydrogen infrastructure—and an up-to-date listing by geographical area—international, Europe, North America, and Pacific Rim.

Barriers and challenges

Only a small number of hydrogen technologies, systems, and components are in operation, and many are in pre-commercial development and are still proprietary. This means that only limited data are available on the operational and safety aspects of these technologies. Historical data used in establishing safety parameters for hydrogen are out-of-date and need to be assessed and validated.

Other possible barriers are lawsuits and insurability, which could affect the commercialization of hydrogen technologies. New technologies that are not yet recognized in codes and standards will have difficulty in obtaining reasonable insurance and might not be approved in some cases. Also, a general lack of understanding of hydrogen safety exists among local government agencies, fire officials, and the general public. Hydrogen safety competence must be developed, especially for local responders to emergencies.

International code development is often complicated and difficult to achieve because of competitiveness and licensing issues. Individual governments have a limited role in the development of global standards. Inadequate representation by government and industry at international forums leads to difficulties in promoting the findings of international technical committees to industry experts in individual countries. Excessively large footprints are required for hydrogen storage at fueling stations, due to a lack of technical data.

NRC pointed out in its report that in the future, safety issues are most likely to arise when hydrogen is used by consumers who have no special training or exposure to the discipline of industrial procedures. Three areas were pinpointed for special attention in vehicle applications: the fueling process, garage storage, and driving in tunnels. NRC recommended research on hydrogen leakage in enclosed structures, development of low-cost, reliable leak sensors, an extensive physical testing program in advance of technology commercialization, training of local fire and rescue officials, and public education.

The *Hindenburg* myth and other misperceptions

In the 1937 Hindenburg accident, the German airship caught fire while landing, and 37 passengers were killed. Hydrogen was used to keep the airship buoyant. At the time of the accident, the practice was to allow storage of hydrogen inside highly combustible materials, which is not permitted under today's safety regulations.

Hydrogen was widely blamed for the disaster at the time, but research since then has shown that hydrogen did not cause the accident, and none of the Hindenburg fatalities resulted from hydrogen flames. The airship's outer shell was coated with a compound similar to what is now used as rocket fuel. As the airship approached its dock, an electrical charge (probably static electricity) ignited this highly flammable coating and started burning the cloth fabric skin.

Once this fire spread, the diesel fuel onboard the airship burned two victims, and the other 35 people died from jumping or falling to the ground. The hydrogen burned rapidly upward, away from the people inside. All of those who rode the Hindenburg to the ground survived.

Another disaster commonly associated with hydrogen was the Challenger Space Shuttle in 1986. However, experts agree that the catastrophe was not caused by hydrogen. Many people also assume that hydrogen explodes or detonates like a nuclear bomb. But burning hydrogen is a chemical reaction, not a nuclear reaction. The heat needed to set off a hydrogen bomb requires a uranium or plutonium fission bomb. This triggers the thermonuclear fusion of hydrogen's tritium isotope, not the common form of hydrogen.

References

1. Powers, Laurie, "Flexibly Fueled Storage Tank Brings Hydrogen-Powered Cars Closer to Reality," Lawrence Livermore National Laboratory, June 2003, www.llnl.gov.

2. Vaitheeswaran, Vijay V., *Power to the People: How the Coming Energy Revolution Will Transform an Industry, Change Our Lives, and Maybe Even Save the Planet*. Farrar, Straus and Giroux: New York, 2003. ISBN 0-374-23675-5, page 245.

3. Smalley, Richard, "The Future of Fuel: Advances in Hydrogen Fuel Cell Technology," PBS Online NewsHour, October 20, 2003, www.pbs.org.

4. National Research Council, *The Hydrogen Economy: Opportunities, Costs, Barriers, and R&D Needs*. The National Academies Press, 2004 (394 pages). ISBN 0-309-09163-2, www.nap.edu.

5. Vincent, Bill, "Hydrogen and the Law: Safety and Liability," Breakthrough Technologies Institute, Presentation to George Washington University Law School, June 11, 2004.

6. Ibid.

7. Ibid.

8. Rocky Mountain Institute, www.rmi.org.

9. Parsons Brinckerhoff, *Support Facilities for Hydrogen Fueled Vehicles: Conceptual Design and Cost Analysis Study*. California Fuel Cell Partnership, July 2004, www.cafcp.org.

Transportation and Distribution

Today we don't need to think twice about how we get our gasoline, heating oil, propane, or natural gas. You fill up your car at the most convenient gasoline station, or maybe drive a little farther to a cheaper one. Whatever fuel you use for home heating is delivered with no fuss—natural gas via pipeline, heating oil or propane by truck. If your home's fuel tank is outside, you don't even need an appointment for it to be topped up. The bill will be in the mail.

Transportation and distribution of hydrogen isn't that easy, and a lot of new equipment will have to be installed before hydrogen becomes convenient to use. This infrastructure will consist mainly of vehicle fueling stations in the near term, when onsite production units will supply most of our hydrogen.

Later on—as vehicles and micropower plants proliferate and demand for hydrogen grows—we might need to build regional hydrogen distribution hubs and possibly special pipelines if our supplies come from central hydrogen production plants.

Unlike liquid fuels, the cost of delivering hydrogen to dispersed customers is extremely high—roughly five times more expensive than producing the hydrogen, according to Daniel Sperling and Joan Ogden in *Issues in Science and Technology*. Even in an optimistic scenario, the hydrogen delivery process would cost about the same as hydrogen production. That means we'll have to think of new ways to transport and distribute hydrogen, not just mimic existing systems for conventional fuels.[1]

Fortunately, hydrogen's versatility will allow us to choose from a multitude of delivery pathways between production and consumption. Transportation and distribution strategies would evolve tailored to regional and local resources, population, and geographic factors, as well as hydrogen demand patterns and production capabilities.

For example, instead of gasifying biomass into hydrogen, the biomass might be converted to ethanol and trucked to vehicle fueling stations, where it could be reformed into gaseous hydrogen. Some day, liquid hydrogen might even be delivered along with electric power via superconducting cables, using the supercold hydrogen to keep the cables from overheating.

Problems of Hydrogen Transport

Hydrogen in its molecular form is a uniquely difficult commodity to ship in high volumes, for many of the same reasons that complicate hydrogen storage (see chapter 6). On a weight basis, hydrogen contains almost three times as much energy as gasoline, but on a volume basis, the situation is more than reversed.

Compressed hydrogen at 5,000 psi contains 3 megajoules of energy per liter and as a cryogenic liquid, 8 megajoules. Both compare unfavorably to gasoline's energy content of 32 megajoules per liter—10 times that of compressed gas and 4 times that of liquid hydrogen.[2] Also, the energy required to compress or liquefy hydrogen cannibalizes its energy content.

Road shipment

Most hydrogen today is consumed onsite where it's produced near refineries (Fig. 7–1) and chemicals manufacturing plants, so the transportation and distribution network in industrial countries is negligible, compared to oil and natural gas delivery systems. Industrial gas companies who sell merchant hydrogen do move some volumes around, mainly on trucks and through dedicated pipelines.

Fig. 7–1. Hydrogen plant at California refineries. Hydrogen produced at this plant is used in nearby refineries to convert crude oil into clean-burning gasoline and low-sulfur diesel fuel. Photo courtesy of Air Products.

In road transport, trucks can carry compressed gas in loads of cylinders or in a single large tube trailer. Hydrogen pressures can range as low as 100 psi and up to 5,000 psi. Compressed gas trucks and trailers typically travel only 100 to 200 miles from the production facility (200 to 300 km). For longer distances, up to 1,000 miles (1,600 km), cryogenic tankers carry liquid hydrogen. Some liquid hydrogen is also shipped by rail or barge. The liquid fuel is vaporized at the customer site.

As practiced today, none of these methods is cost-effective for delivering enough hydrogen to support widespread use in the future. The energy consumption of compressed gas trucking in particular is unacceptably high at almost any distance. Although this option might be needed to deliver limited volumes in the early years of a transition toward hydrogen, fuel price subsidies would probably be required for hydrogen-powered vehicles to compete, said the U.S. National Research Council (NRC) in its report, *The Hydrogen Economy: Opportunities, Costs, Barriers, and R&D Needs.*[3]

Hydrogen produced onsite at fueling stations and possibly at homes and businesses would be more expensive than hydrogen from central plants. However, transportation and distribution costs would make onsite production the logical choice for the foreseeable future.

Hydrogen pipelines

Hydrogen delivery by pipeline is the lowest cost option for high volumes. However, existing hydrogen pipelines owned by merchant vendors are limited to areas with a large concentration of refineries or chemical plants. In the United States, these centers are in Indiana, California, Texas, and Louisiana and have a total of about 700 miles of hydrogen pipelines (1,100 km). This network is far too small to support widespread hydrogen use.

The world's first long-distance hydrogen pipeline, opened in 1938, was a 150-mile (240-km) network in Germany, which is still in operation serving several chemical plants. Northern Europe is home to several clusters of refineries and chemicals complexes, with hydrogen pipelines in the region totaling around 900 miles (1,400 km).[4]

Some natural gas pipelines could be switched to deliver hydrogen—several already have been converted and are in operation—but many existing gas lines might not be suitable. For example, some natural gas transmission lines use high-grade steel that allows a thinner pipe wall.

Exposure to hydrogen would embrittle this kind of pipe and fittings, resulting in leaks, cracks, and failures.

The severity of the embrittlement problem depends on the type of steel and weld used and the pressure in the pipeline. Technology is available to prevent embrittlement, but depending on the design, distribution costs could be affected. Probably, each pipeline segment would have to be considered on a case-by-case basis. Some steels and welds would be compatible, while others might be subject to embrittlement, particularly the welds in older segments. Reciprocating compressors would generally have to be refitted with new seals and valves, while centrifugal compressors would not work on hydrogen at all and would have to be replaced entirely. Both types are common in the natural gas pipeline system.[5]

Also, hydrogen's molecules are smaller than natural gas and would escape more easily from conventional pipelines, and lubricating oils used in the pumps that propel natural gas through the line could be incompatible with hydrogen.

For building new pipelines, hydrogen's properties would require different materials and welding techniques. Different types of steel pipe would be needed to mitigate the embrittlement problem, as well as more costly valve seal connections and other pipeline components.

For cost-effective delivery, hydrogen pipelines would have to be designed for operation at higher pressures and flow velocities than natural gas pipelines, roughly three times as high for the same diameter. Alternatively, hydrogen pipelines would have to be 50% larger in diameter than natural gas lines to achieve the same energy transmission rate, making them more expensive, said NRC.

Argonne National Laboratory compared the capital cost of building small-diameter natural gas and hydrogen pipelines. Hydrogen pipelines could cost 75% more, or even double the amount in the smallest diameter (Table 7–1).

TABLE 7–1. PIPELINE CONSTRUCTION COSTS

Pipe diameter, inches	Capital cost, $/mile	
	Natural gas	Hydrogen
3	200,000	400,000
9	500,000	900,000
12	600,000	1,000,000
14	800,000	1,400,000

Source: Argonne National Laboratory[6]

The energy required to move hydrogen through a pipeline is estimated to be 3.8 to 4.6 times greater than to move natural gas containing the same amount of energy. Also, some of the hydrogen would be consumed along the way at pumping stations, as is the practice with most natural gas pipelines.[7]

This compression energy consumption is estimated at 1.4% of the hydrogen for every 100 miles (160 km) of pipeline distance. For example, in the case of a pipeline 2,000 to 2,700 miles long (3,200 to 4,300 km), only 60% to 70% of the hydrogen would reach its destination.

Another analysis concludes that hydrogen transmission costs might be about 50% higher than for natural gas, because the capacity of the compressor stations on an energy basis is about one-third less. In a pipeline system optimized to carry hydrogen, the pipe's dimensions and the size and spacing of the compressors would have to be changed to accommodate these factors.[8]

Transportation economics

Hydrogen can be produced at a central plant today at a cost ranging from $0.96 to $1.22/kg, depending on the feedstock (natural gas or coal) and whether or not carbon is sequestered. Strikingly, pipeline shipment and dispensing of hydrogen adds an estimated cost of $0.96/kg, essentially the same as the production cost, said NRC. Even with future technology improvements, shipping and distribution costs will be much higher than those for our existing gasoline system, at 19¢/gallon (a gallon of gasoline is the energy equivalent of about a kilogram of hydrogen).

In the early stages of building up a hydrogen infrastructure, the transportation of cryogenic liquid hydrogen via trucks or rail could play a significant role, said NRC, despite the expense of the process. Over-the-road shipment of liquid hydrogen and dispensing at a vehicle fueling station is estimated to cost any where from $1.40 to $2.42/kg, on top of hydrogen production costs. More efficient liquefaction technology might shave some of these costs.

Outlook

Most analysts agree that an extensive, long-distance, high-volume pipeline network for hydrogen will not be needed for a long time. Even though pipelines are the cheapest way to move large volumes of hydrogen, the capital cost of building such a network could be prohibitive, even when spread over many years. For cities with dense populations, dedicated pipelines in strategic locations might make sense economically, once a substantial number of vehicles are running on hydrogen.

In most other situations, however, onsite hydrogen production is much less capital-intensive than pipeline construction and undoubtedly more economical than over-the-road shipping, said NRC. Onsite hydrogen production is expected to dominate the marketplace through at least 2020 and possibly for another 10 to 15 years beyond that.

Researchers are developing advanced hydrogen storage technologies that use hydrides and other chemicals to store hydrogen atoms at low pressures (see chapter 6). If lightweight systems are perfected for onboard vehicle storage, these techniques might improve the economics of road shipment as well, since chemical storage containers might be transported by conventional trucks rather than requiring high-pressure or cryogenic containers. Chemical storage could even transform the way we produce and deliver hydrogen in the future, once its potential is explored more thoroughly.

NRC envisions a gradual evolution in the transportation and delivery of hydrogen. Transitions will occur several times as hydrogen demand increases over many decades. This will involve continuous, overlapping shifts from small-scale delivery and onsite production toward centralized production with a network of regional distribution hubs, intermediate storage facilities, and some dedicated pipelines.

This vision of a creative, evolutionary approach is shared by Sperling and Ogden, who suggest that in the future, the transportation fuels industry might become more closely linked with the electric power and chemical industries. The economics of hydrogen production might be improved by co-producing electricity and chemical products, spurring synergies among these businesses.[9]

For example, scientists have proposed a concept called the nuclear-hydrogen SuperGrid or energy superhighway. High-capacity, long-distance direct-current power lines made of superconducting cable would be cooled by liquid hydrogen produced at the power plant. Some of the hydrogen would be delivered at the end point for use in vehicles and micropower plants.

Fueling Station Infrastructure

Building a network of hydrogen fueling stations will be critical to the success of hydrogen as a transportation fuel. No one wants to drive a new car out of the showroom without confidence that fuel will be available when and where they need it.

In Asia and Europe, where countries are smaller and driving distances are relatively shorter, the convenience of filling up a fuel cell vehicle could be easier to achieve than in the United States. The 25 countries in the extended European Union are estimated to have about 230 million passenger cars and commercial vehicles and more than 100,000 filling stations. In Germany alone, clusters of cheap hydrogen sources from the chemical industry total about 36 billion cubic feet per year (1 billion nm^3/yr), which could support about 600 fueling stations.[10]

Americans are unlikely to accept a technology that requires them to go very far out of their way, and some of the distances between population centers are vast compared to other industrial nations. Currently, the United States has more than 170,000 gasoline stations serving more than 240 million vehicles. U.S. drivers logged more than 2.57 trillion miles in passenger cars alone in 2001, and the total for all vehicles was more than 4.4 trillion miles.[11]

Many observers have characterized development of the hydrogen fueling infrastructure as a classic chicken-and-egg dilemma. People won't buy fuel cell vehicles that can't be refueled conveniently, but industry won't build fueling stations until enough hydrogen vehicles are on the road. Scientists have pegged the cost of constructing a national U.S. network of hydrogen fueling stations at up to $400 billion for a fleet of 100 million vehicles. The entire production and delivery infrastructure could cost more than $600 billion.[12]

However, the problem might not be as insurmountable as it seems at first glance. For one thing, we don't need to build up all of our fueling stations from scratch. In many locations, hydrogen storage tanks and dispensing pumps can be added to existing gasoline stations.

Also, thousands of fleet vehicles have their own refueling facilities where adding hydrogen service is easier than at a congested public station. And technology developers are coming up with clever ways to package hydrogen equipment in modules that can be integrated with existing facilities or combined into new stations.

Even more important, oil companies and automotive manufacturers are backing the switchover. Multinationals BP and Shell are installing hydrogen fueling stations worldwide, as part of fuel cell vehicle demonstrations by virtually every major automaker (see chapter 4 and Appendices A–C). A total of at least 80 stations have been built or are under development, including some at research facilities.

Infrastructure cost estimates

Surprisingly, the energy and auto industries believe that a hydrogen fueling infrastructure can be initiated, at least, for a reasonable investment. General Motors estimated that an early network in the United States would cost $10 to $15 billion. This would pay for 11,700 new fueling stations—enough that drivers in urban areas would always be within 2 miles of a hydrogen station (3 km), and major highways would have a station every 25 miles (40 km). This concentration would support about a million fuel cell vehicles.[13]

As GM's head of research Larry Burns said, "Fuel infrastructure will not be the show-stopper for fuel cells. This is quite doable."[14]

In a similar analysis, Shell estimated that an initial fueling station infrastructure would cost about $19 billion. This amount would allow retrofitting of about 44,000 existing stations (one-fourth of the U.S. total) with hydrogen dispensing pumps. Shell anticipated that retrofits would cost about $420,000 per station.

Shell's early experience opening hydrogen service at an existing station in Washington, DC, suggests that this cost estimate might be somewhat low. However, new codes and standards that facilitate the installation of hydrogen fueling equipment are expected to alleviate some requirements and costs.

Although the automotive and energy industries have yet to request government subsidies for building up a fueling station infrastructure, tax breaks and other financial incentives probably will be needed to get a massive effort under way. GM has compared the undertaking to America's investment in railroads in the 19th century and in the interstate-highway network in the 20th century.

Germany's L-B-Systemtechnik GmbH analyzed the investment costs for a European hydrogen supply and refueling infrastructure in the 2010–20 time frame, assuming that half the hydrogen would be produced from natural gas and half from wind-powered electrolysis. Compressed gas fueling stations were estimated to cost about 3.7 million euros each, including hydrogen production. A liquid hydrogen fueling station would cost about 4.4 million euros.

Building 15,000 European fueling stations by 2010 would entail a cumulative investment of about 60 billion euros, rising to 241 billion euros to reach 60,000 stations in 2020. Of these totals, the refueling stations account for about 33% and hydrogen production for the other 67%.

A similar analysis by the European Union's HyNet group estimated the investment cost of 5,000 fueling stations at 3.8 billion euros for compressed gas service based on onsite electrolysis and 7 billion euros for centrally produced liquid hydrogen from natural gas. Corresponding estimates for 10,000 stations were 7 billion and 13.8 billion euros, respectively. HyNet is working on a European Hydrogen Energy Roadmap that includes infrastructure development. The International Partnership for a Hydrogen Economy also is researching methods of transporting and distributing hydrogen.

Fueling station equipment

Four major components are needed for a compressed gas hydrogen fueling station—hydrogen production, compression, storage, and fuel dispensing. In most fueling stations, hydrogen would be produced onsite by reforming natural gas (or other hydrocarbon fuel) or electrolyzing water. Hydrogen fuel could be trucked or piped in if the station is close to a refinery or chemicals complex, eliminating the need for onsite production.

Compressors boost the hydrogen's pressure to levels required for gaseous storage, usually around 6,000 psig. High-pressure compressed gas cylinders store enough hydrogen at the station to meet expected vehicle fueling demand, which changes throughout the day.

Recent updates to international fire codes allow compressed gas cylinders to be installed on top of service station canopies. Overhead placement prevents vehicles from colliding with the tanks and also would allow the tanks to vent safely upward in case of an emergency. Remotely activated discharge devices control the emergency release of hydrogen.

Fuel pumps and nozzles dispense the hydrogen into vehicles' onboard tanks, which carry it at pressures of 3,600 or 5,000 psig. Future onboard storage tanks could operate at 10,000 psig. The refueling process takes about 4 minutes, comparable to a gasoline fill-up.

Hydrogen dispensing nozzles connect more tightly to the vehicle than gasoline pump nozzles. Also, a communication cable connects to the vehicle and sends information to a computer about the vehicle's onboard tank conditions to verify that the tank is safe and ready for refueling. Hydrogen dispensing nozzles are expensive, but manufacturers are working to bring costs down, and some fuel cell vehicle programs subsidize the cost of testing nozzles to industry standards.

If liquid hydrogen is being trucked in, the fueling station would also incorporate a cryogenic storage tank and a vaporizer that converts the liquid to gaseous hydrogen. International fire codes now allow cryogenic hydrogen storage underground, protecting the tank from vehicles and possible vandalism, as well as allowing safe venting aboveground.

To test various fueling station designs in the United States, the Department of Energy is funding a five-year program that will build 24 stations in California. One will be located alongside a pipeline, and moveable equipment will be developed that can be placed at a series of existing gasoline service stations. Some of the fueling stations will dispense both gaseous and liquid hydrogen. Up to 65 fuel cell vehicles supplied by Toyota, Nissan, and Honda will use the stations as part of the demonstration, plus 15 hydrogen engine-powered vehicles provided by BMW.

Automated station at Munich airport. Linde AG, Europe's leading supplier of industrial gases, participated in the installation of a hydrogen fueling station at the Munich airport that provides both gaseous and liquid hydrogen (Fig. 7–2). The public station, opened in 1999, features a robotic arm that fully automates the liquid refueling process.

Fig. 7–2. Munich airport hydrogen fueling station. At the Munich airport fueling station, a robotic arm automatically fills the vehicle with liquid hydrogen. Photo courtesy of Linde.

Drivers simply swipe a magnetic-stripe ID card into a reader to begin refueling, without leaving the car. The arm swings out, locks onto the car's fueling port, and fills the tank with about 120 liters of liquid hydrogen in about 90 seconds.[15]

Linde also built the world's first hydrogen fueling station using 10,000-psi (700-bar) compressed gas technology. The higher energy density afforded by the increased pressure extends the travel range to 250 miles (400 km). The station is located at the Opel test center near Offenbach, Germany.

West Sacramento facility. This demonstration station, which opened in 2000, is located at the headquarters of the California Fuel Cell Partnership. BP and Shell contributed to its design, along with ChevronTexaco, Exxon-Mobil, and industrial gas suppliers Air Products and Praxair.

The fueling station imports liquid hydrogen and stores it in a double-wall, 4,500-gallon tank that operates at 50 psig. The cryogenic tank was designed for 150 psig and has been tested up to 225 psig, so it could contain additional hydrogen if needed.

After the liquid fuel is vaporized, a compressor boosts its pressure to 6,250 psi for storage in three compressed gas tubes, which have relief valves in case of overpressure. The tubes were designed for up to 7,255 psig and tested to 1.5 times that level.

Each of three fuel dispensing pumps provides hydrogen at a different condition: liquid, 3,600-psi gas, and 5,000-psi gas. The liquid dispensing line, which siphons directly off of the liquid storage tank, is vacuum-jacketed to keep the fuel cold. It delivers a few liquid gallons per minute.

The two compressed gas dispensers fill up a vehicle's hydrogen tank in about 4 minutes. They have several safety features, including overpressure protection, remote shutoff capability, a breakaway connector in case the car is driven away while the hose is attached, and non-interchangeable nozzles to prevent inadvertent use of the wrong hydrogen pressure. A combination ultraviolet-infrared detector (UV-IR) with an alarm monitors the station for hydrogen leaks or fires.

Air Products also built the natural gas reforming system that provides hydrogen to a fueling station in Las Vegas—the world's first to co-produce both hydrogen and electricity onsite (Fig. 7–3). The station dispenses compressed hydrogen and a hydrogen-natural gas blended fuel, while a Plug Power PEM hydrogen fuel cell exports electric power to the grid.

FIG. 7–3. LAS VEGAS HYDROGEN FUELING STATION. THIS DISPENSING PUMP
PROVIDES COMPRESSED HYDROGEN TO FUEL CELL VEHICLES AT THE LAS VEGAS
STATION, WHICH ALSO GENERATES POWER FROM A STATIONARY PEM FUEL CELL.
PHOTO COURTESY OF AIR PRODUCTS.

Electrolysis-based fueling stations. Several manufacturers offer commercial electrolyzers suitable for hydrogen fueling stations (see chapter 5). One leader is Stuart Energy Systems, which provided electrolysis units for four stations serving Europe's CUTE fuel cell buses as well as several other locations. One advantage of hydrogen based on electrolysis is that wind turbines or solar energy can supply the fueling station with electric power for water-splitting, avoiding any carbon emissions in the hydrogen supply chain.

Stuart's hydrogen fueling products are designed as four modules—hydrogen generation, compression, storage, and fuel dispensing—that can serve as individual units or can be combined to form an integrated fueling station. The product line can be scaled to a wide range of sizes, and Stuart is working with Shell to design a HomeFueler® system for residential use (see chapter 4).

At a fueling station in Malmo, Sweden, Stuart Energy provided the electrolytic hydrogen generator, compressor, storage unit, and fuel dispenser. The system provides Malmo's bus fleet with 75 to 80 kg/d of hydrogen (1,368 scf/h, or 36 nm^3/h). The Malmo station's storage module contains 95 kg of hydrogen at 5,600 psi (393 bar).

European utility Sydkraft also supplies natural gas bus fuel at the station. For environmental reasons, many European buses were previously converted from diesel to natural gas, and researchers are working on engines that can use blends of hydrogen and natural gas fuel. The Stuart dispenser incorporates a hydrogen-natural gas mixing system that can supply a range of fuel blends. Pure hydrogen is delivered at 5,000 psi (350 bar) and hydrogen-natural gas blends at 3,600 psi (250 bar).

Electrolyzers will also provide hydrogen at two vehicle fueling stations in California. Proton Energy Systems' HOGEN 200 units are being integrated with Air Products' Series 200 storage and dispensing equipment, offering customers flexibility in using hydrogen generated onsite or delivered hydrogen.

Outlook

Onsite natural gas reforming and water electrolysis will probably supply hydrogen fueling stations for the foreseeable future, except in locations near refineries or chemical plants. These two onsite hydrogen production technologies avoid the high costs of transporting and distributing gaseous or liquid hydrogen.

Onsite production also takes advantage of the existing infrastructure for delivering natural gas, electricity, and water in industrial nations. Fueling station developers are working on ways to integrate onsite electrolytic hydrogen production with renewable energy sources, which would not only avoid hydrogen delivery costs but also reduce air pollution.

References

1. Sperling, Daniel, and Ogden, Joan, "The Hope for Hydrogen," *Issues in Science and Technology*, Spring 2004, page 85.

2. National Research Council, *The Hydrogen Economy: Opportunities, Costs, Barriers, and R&D Needs.* The National Academies Press, 2004 (394 pages). ISBN 0-309-09163-2, www.nap.edu.

3. Ibid.

4. Wurster, Reinhold, "Hydrogen Road Vehicles and their Refuelling Infrastructure: Strategic Considerations and The European Situation in Regulations & Standards," L-B-Systemtechnik GmbH, presentation to the H_2IT Associazione italiana idrogeno e celle a combustibile, Milan, April 16, 2004.

5. Morgan, Daniel, and Sissine, Fred, "Hydrogen: Technology and Policy," CRS Report for Congress No. 95-540 SPR, April 28, 1995, www.ncseonline.org.

6. Mintz, Marianne, Folga, Stephen, Molburg, John, and Gillette, Jerry, "Cost of Some Hydrogen Fuel Infrastructure Options," Argonne National Laboratory, January 16, 2002, www.anl.gov.

7. Eliasson, Baldur, and Bossel, Ulf, "The Future of the Hydrogen Economy: Bright or Bleak?" Paper presented at Fuel Cell World, Lucerne, July 5, 2000.

8. Morgan, Daniel, and Sissine, Fred, "Hydrogen: Technology and Policy," CRS Report for Congress No. 95-540 SPR, April 28, 1995, www.ncseonline.org.

9. Sperling, Daniel, and Ogden, Joan, "The Hope for Hydrogen," *Issues in Science and Technology*, Spring 2004, page 85.

10. Wurster, Reinhold, "Hydrogen Road Vehicles and their Refuelling Infrastructure: Strategic Considerations and The European Situation in Regulations & Standards," L-B-Systemtechnik GmbH, presentation to the H_2IT Associazione italiana idrogeno e celle a combustibile, Milan, April 16, 2004.

11. Bureau of Transportation Statistics, www.bts.gov.

12. Mintz, Marianne, Folga, Stephen, Molburg, John, and Gillette, Jerry, "Cost of Some Hydrogen Fuel Infrastructure Options," Argonne National Laboratory, January 16, 2002, www.anl.gov.

13. Ball, Jeffrey, "Industries Seek Fuel-Cell Funding," *The Wall Street Journal*, November 18, 2003.

14. "Fuel Cells: A new kind of gas station," *The Economist*, December 6, 2003, page 59.

15. Hoffman, Peter, *Tomorrow's Energy: Hydrogen, Fuel Cells, and the Prospects for a Cleaner Planet.* The MIT Press: Cambridge, MA, 2001. ISBN 0-262-08295-0, page 123.

appendix

Fuel Cell Light
Duty Vehicles

Table created by Fuel Cells 2000 and US Fuel Cell Council
(available for downloading at www.fuelcells.org/info/charts/carchart.pdf)

Automaker	Vehicle Type	Year Shown	Engine Type	Fuel Cell Size/type	Fuel Cell Mfr.	Range (mi/km)	MPG Equivalent*	Max. Speed	Fuel Type	Commercial Intro.	Picture
BMW	Series 7 (745 h) (Sedan)	2000	ICE (fuel cell APU)	5kW/PEM	UTC	180mi 300km	N/a	140 mph	Gasoline/Liquid hydrogen	Limited intro in 2000 (Munich Airport Hydrogen Vehicle Project)	
Daihatsu	MOVE EV - FC (micro van)	1999	Fuel cell/battery hybrid	16kW/PEM	Toyota	N/a	N/a	N/a	Methanol		
	MOVE FCV - K II (mini vehicle)	2001	Fuel cell/battery hybrid	30 kW/PEM	Toyota	75mi 120km	N/a	65mph 105km/h	Compress. hydrogen @ 3600 psi	Japan road testing started in early 2003.	
Daimler-Chrysler	NECAR 1 (180 van)	1994	12 fuel cell stacks	50kW/PEM	Ballard	81mi 130km	N/a	56mph 90km/h	Compress. hydrogen @ 4300 psi		
	NECAR 2 (V-Class)	1996	Fuel cell	50kW/PEM	Ballard	155mi 250km	N/a	68mph 110km/h	Compress. hydrogen @ 3600 psi		
	NECAR 3 (A-Class)	1997	2 fuel cell stacks	50kW/PEM	Ballard Mark 700 Series	250mi 400km	N/a	75mph 120km/h	10.5 gal. of Liquid methanol	First methanol reforming FCV	
	NECAR 4 (A-Class)	1999	Fuel cell	70kW/PEM	Ballard Mark 900 Series	280mi 450km	N/a	90mph 145km/h	Liquid hydrogen		
	Jeep Commander 2 (SUV)	2000	Fuel cell/(90 kW)/battery hybrid	50kW/PEM	Ballard Mark 700 Series	118mi 190km	24 mpg (gasoline equiv.)	N/a	Methanol	Jeep Commander 1 came out in 1999.	

Vehicle	Year	Type	Power	Stack	Range	Economy	Speed	Fuel	Notes
NECAR 4 - Advanced (California NECAR)	2000	Fuel cell	85kW/PEM	Ballard Mark 900 Series	124mi 200km	53.46 mpg equiv. (CaFCP est.)	90mph 145 km/h	4 lbs. (1.8kg) of Compress. hydrogen @ 5,000 psi	
NECAR 5 (A-class)	2000	Fuel cell	85kW/PEM	Ballard Mark 900 Series	280mi 450km	N/a	95mph 150km/h	Methanol	
DMFC go-cart (one-person vehicle)	2000	Fuel cell	3kW/DMFC	Ballard Mark 900 Series	9.3mi 15km	N/a	22mph 35km/h	Methanol (directly)	6kW DMFC built by DC and Ballard is largest in world
NECAR 5.2 (A-class)	2001	Fuel cell/battery hybrid	85kW/PEM	Ballard Mark 900 Series	300mi 482km	N/a	95mph 150km/h	Methanol	Awarded a road permit for Japanese roads. Completed CA – DC drive.
Sprinter (van)	2001	Fuel cell	85kW/PEM	Ballard Mark 900 Series	93mi 150km	N/a	75mph 120km/h	Compress. Hydrogen @ 5,000 psi	Delivered to Hamburg parcel service, Hermes as part of the W.E.I.T. hydrogen project
Natrium (Town & Country Mini Van)	2001	Fuel cell/(40 kW) battery hybrid	54kW/PEM	Ballard Mark 900 Series	300mi 483km	30 mpg equiv.	80mph 129km/h	Catalyzed chemical hydride - Sodium Borohydride	Uses Millennium Cell's 'Hydrogen on Demand' system with a 53 gallon fuel tank
F-Cell (A-class)	2002	Fuel cell/battery hybrid	85kW/PEM	Ballard Mark 900 Series	90mi 145km	56 mpg equiv.	87mph 140km/h	4 lbs. (1.8kg) of Compress. hydrogen @ 5,000 psi	60 fleet vehicles in US, Japan, Singapore, and Europe starting in 2003 – small fleet in Michigan operated by UPS.

Manufacturer	Model	Year	Type	Power	Fuel Cell	Range	Efficiency	Speed	Fuel	Notes
	Jeep Treo	2003	Fuel cell	N/a	N/a	N/a	N/a	N/a	N/a	Unveiled at Tokyo Motor Show – drive by wire technology
ESORO	Hycar	2001	Fuel cell/ battery hybrid	6.4kW/ PEM	N/a	224mi 360km	N/a	75mph 120km/h	Compress. Hydrogen	Switzerland's first FCV
Fiat	Seicento Elettra H2 Fuel Cell	2001	Fuel cell/ battery hybrid	7kW/ PEM	Nuvera	100mi 140km	N/a	60mph 100km/h	Compress. hydrogen	Next generation due in 2003 w/ Nuvera fuel cells
	Seicento Elettra H2 Fuel Cell	2003	Fuel cell/ battery hybrid	N/a	Nuvera	N/a	N/a	N/a	Compress. hydogen	Being investigated for use in Milan, Italy, where gasoline and diesel fueled vehicles are banned on smoggy days.
Ford Motor Company	P2000 HFC (sedan)	1999	Fuel cell	75kW/ PEM	Ballard Mark 700 Series	100mi 160km	67.11 mpg equiv. (CaFCP est.)	N/a	Compress. hydrogen	First FCV by Ford
	Focus FCV	2000	Fuel cell	85kW/ PEM	Ballard Mark 900 Series	100mi 160km	N/a	80mph 128km/h	Compress. hydrogen @ 3,600 psi	
	THINK FC5	2000	Fuel cell	85kW/ PEM	Ballard Mark 900 Series	N/a	N/a	80mph/ 128km/h	Methanol	
	Advanced Focus FCV	2002	Fuel cell/ battery hybrid	85kW/ PEM	Ballard Mark 900 Series	180mi 290km	~50 mpg equiv.	N/a	8.8 lb. (4kg) Compress. H2 @ 5,000 psi	~40 fleet vehicles introduction Germany, Vancouver & CA in 2004
	GloCar Concept Only	2003	Fuel Cell	N/a	N/a	N/a	N/a	N/a	N/a	Powered by fuel cells, it uses LED lights to change body panel colors, intensity, and frequency.

Manufacturer	Model	Year	Type	Power	FC	Range	Fuel economy	Speed	Fuel	Comments
General Motors/ Opel	EV1 FCEV	1997	Fuel cell/ battery hybrid	N/a	N/a	N/a	N/a	N/a	Methanol	0-60 mph in 9 seconds
	Sintra (mini-van)	1997	Fuel cell	50kW/ PEM	N/a	N/a	N/a	N/a	N/a	Wants to be 1st automaker to sell 1 million FCVs profitably
**Hydrogenic s works with GM on FC development	Zafira (mini-van)	1998	Fuel cell	50kW/ PEM	Ballard	300mi 483km	80 mpg equiv.	75mph 120km/h	Methanol	GM has ceased efforts regarding methanol (2001)
	Precept FCEV Concept only	2000	Fuel cell/ battery hybrid	100kW/ PEM	GM**	500mi 800km (est.)	108 mpg equiv. (est.)	120mph 193km/h	Hydrogen (stored in metal hydride)	These are concept projections
	HydroGen 1 (Zafira van)	2000	Fuel cell/ battery hybrid	80kW/ PEM	GM**	250mi 400km	N/a	90mph 140km/h	16 gal. of Liquid hydrogen	GM plans to sell 75kW hydrogen stationary fuel cell generators in 2005
	HydroGen 3 (Zafira van)	2001	Fuel cell	94kW/ PEM	GM**	250mi 400km	N/a	100mph 160km/h	Liquid hydrogen	Being used by FedEx Corp. in Tokyo, Japan from 6/2003 – 6/2004
	Chevy S-10 (pickup truck)	2001	Fuel cell/ battery hybrid	25kW/ PEM	GM**	240mi 386km	40 mpg	70 mph	Low sulfur, clean gasoline (CHF)	GM has partnership with Toyota on reforming
	AUTOnomy Concept only	2002	Fuel cell	N/a	N/a	N/a	Projected 100 mpg	N/a	N/a	GM's 2010 FCV concept Freedom of Design
	Hy-Wire Proof of Concept	2002	Fuel cell	94kW/ PEM	GM**	80mi 129km	~41 mpg (gas equiv.)	97mph 160km/h	4.4 lbs.(2kg) Compress. h2 @ 5,000 psi	Uses HydroGen3's powertrain, so range & mpg theoretically could = HydroGen3

	Model	Year	Type	Power	Stack	Range	Efficiency	Speed	Storage	Notes
GM (Shanghai) PATAC	Advanced HydroGen 3 (Zafira van)	2002	Fuel cell	94kW/ PEM	GM**	170mi 270km	~55 mpg (gas equiv.)	~100mph 160km/h	6.8lbs. (3.1kg) Compress. h2 @ 10,000 psi	1st FCV to incorporate 10,000 psi tanks (by Quantum). 6 placed in Washington DC.
	Diesel Hybrid Electric Military truck	2003	Fuel cell APU	5kW/ PEM	Hydrogenics	N/a	N/a	N/a	Low pressure metal hydrides	Turbo diesel ICE/battery hybrid with PEM FC APU. Under eval. for US Army's new fleet of 30,000 light tactical vehicles.
	Phoenix (Mini van)	Oct. 2001	Fuel cell/ battery hybrid	25kW/ PEM	Shanghai GM**	125mi 200km	N/a	70mph 113km/h	Compress. hydrogen	Seventh FCV prototype out of China
Honda	FCX-V1	1999	Fuel cell/ battery hybrid	60kW/ PEM	Ballard Mark 700 Series	110mi 177km	N/a	78mph 130km/h	Hydrogen (stored in metal hydride)	
	FCX-V2	1999	Fuel cell	60kW/ PEM	Honda	N/a	N/a	78mph 130km/h	Methanol	Honda has strict focus on pure hydrogen FCVs (2001)
	FCX-V3	2000	Fuel cell/ Honda ultra capacitors	62kW/ PEM	Ballard Mark 700 Series	108mi 173km	N/a	78mph 130km/h	26 gal. of Compress. hydrogen at 3600 psi	
	FCX-V4	2001	Fuel cell/ Honda ultra capacitors	85kW/ PEM	Ballard Mark 900 series	185mi 300km	~50 mpg (gas equiv.)	84mph 140km/h	130 L (3.75kg) Compress. H2 @ 5,000 psi	Completed Japanese road testing - 1st FCV to receive CARB & EPA emission certs.
	FCX	2002	Fuel cell/ Honda ultra capacitors	85kW/ PEM	Ballard Mark 900 series	220mi 355km	~50 mpg (gas equiv.)	93mph 150km/h	156.6 L Compress. hydrogen @ 5000 psi	LA (5 total) Japan's Cabinet Office (1) leasing at $6500/month each (12/2/02)
	Kiwami concept	2003	Fuel cell	N/a	N/a	N/a	N/a	N/a	Hydrogen	Unveiled at Tokyo Motor Show

Manufacturer	Model	Year	Type	Power	Fuel Cell	Range		Speed	Fuel	Notes
Hyundai	Santa Fe SUV	2000	Ambient-pressure Fuel cell	75kW/PEM	UTC Fuel Cells	100mi 160km	N/a	77mph 124km/h	Compress. hydrogen	
	Santa Fe SUV	2001	Ambient-pressure Fuel cell	75kW/PEM	UTC Fuel Cells	250mi 402km	N/a	N/a	Compress. hydrogen	2003 – 2004 limited intro. to power utilities & research institutes
Mazda	Demio (compact passenger car)	1997	Fuel cell/ultra capacitor hybrid	20kW/PEM	Mazda	106mi 170km	N/a	60mph 90km/h	Hydrogen (stored in metal hydride)	
	Premacy FC-EV	2001	Fuel cell	85kW/PEM	Ballard Mark 900 Series	N/a	N/a	77mph 124km/h	Methanol	Awarded road permit for Japanese roads in 2001 – undergoing public road testing
Mitsubishi	SpaceLiner Concept only	2001	Fuel cell/battery hybrid	40kW/PEM	N/a	N/a	N/a	N/a	Methanol	Commercial target date in 2005
	Grandis FCV (mini-van)	2003	Fuel cell/battery hybrid	68kW/PEM	Daimler Chrysler/Ballard	92mi 150km	N/a	87mph 140km/h	Compress. Hydrogen	Will be launched in Europe in 2004.
Nissan	R'nessa (SUV)	1999	Fuel cell/battery hybrid	10kW/PEM	Ballard Mark 700 Series	N/a	N/a	44mph 70km/h	Methanol	Partnership with Renault for gasoline fueled FCV until 2006
	Xterra (SUV)	2000/2001	Fuel cell/battery hybrid	85kW/PEM	Ballard Mark 900 Series & UTC Fuel Cells**	100mi 161km	N/a	75mph 120km/h	Compress. hydrogen	
**Made prototypes w/ each fuel cell stack	X-TRAIL (SUV)	2002	Fuel cell/battery hybrid	75kW/PEM	UTC Fuel Cells (Ambient-pressure)	N/a	N/a	78mph 125km/h	Compress. hydrogen @ 5,000 psi	Approved for Japanese Public road testing – limited marketing later in 2003

Manufacturer	Vehicle	Year	Type	Power	FC Maker	Range		Speed	Fuel	Notes	
PSA Peugeot Citron	Effis (commuter concept)	2003	Fuel cell/battery hybrid	N/a	N/a	N/a	N/a	N/a	N/a	Unveiled at Tokyo Motor Show	
	Peugeot Hydro-Gen	2001	Fuel cell/battery hybrid	30kW/PEM	Nuvera	186m 300km	N/a	60mph 95km/h	Compress. hydrogen		
	Peugeot Fuel Cell Cab "Taxi PAC"	2001	Fuel cell/battery hybrid	55kW/PEM	H Power	188mi 300km	N/a	60mph 95km/h	80 Liters Compress. hydrogen @ 4300 psi		
	H2O fire-fighting Concept only	2002	Battery/fuel cell APU	N/a	N/a	N/a	N/a	N/a	Catalyzed chemical hydride - Sodium Borohydride	Uses Millennium Cell's 'Hydrogen on Demand' system	
Renault	EU FEVER Project (Laguna wagon)	1997	Fuel cell/battery hybrid	30kW/PEM	Nuvera	250mi 400km	N/a	75mph 120km/h	Liquid hydrogen	Partnership with Nissan on gasoline fueled FCV	
Suzuki	Covie Concept only	2001	Fuel cell	N/a	GM	N/a	N/a	N/a	N/a		
	Mobile Terrace	2003	Fuel cell	N/a	GM	N/a	N/a	N/a	Hydrogen	Unveiled at Tokyo Motor Show	
Toyota	RAV 4 FCEV (SUV)	1996	Fuel cell/battery hybrid	20kW/PEM	Toyota	155mi 250km	N/a	62mph 100km/h	Hydrogen (stored in metal hydride)		
	RAV 4 FCEV (SUV)	1997	Fuel cell/battery hybrid	25kW/PEM	Toyota	310mi 500km	N/a	78mph 125km/h	Methanol		
	FCHV-3 (Kluger V/Highlander SUV)	2001	Fuel cell/battery hybrid	90kW/PEM	Toyota	186mi 300km	N/a	93mph 150km/h	Hydrogen (stored in metal hydride)	Toyota is developing a Japanese residential 1kW stationary fuel cell system for 2005	

Vehicle	Year	Type	Stack	Manufacturer	Range		Speed	Fuel	Notes
FCHV-4 (Kluger V/ Highlander SUV)	2001	Fuel cell/ battery hybrid	90kW/ PEM	Toyota	155mi 250km	N/a	95mph 152km/h	Compress. Hydrogen @ 3,600 psi	Completed Japanese road testing
FCHV-5 (Kluger V/ Highlander SUV)	2001	Fuel cell/ battery hybrid	90kW/ PEM	Toyota	N/a	N/a	N/a	Low sulfur, clean gasoline (CHF)	Partnered with GM on gasoline CHP reforming technology
FCHV (Kluger V/ Highlander SUV)	2002	Fuel cell/ battery hybrid	90kW/ PEM [122 hp]	Toyota	180mi 290km	N/a	96mph 155km/h	Compress. hydrogen @ 5,000 psi	3 leased to UC Irvine, 3 to UC Davis & 4 to Japanese gov't agencies (12/2/02) for 30 months at $10,000/month each, 6 more to be leased to Japanese local gov'ts and private co.'s
FINE-S Concept only	2003	Fuel cell	N/a	N/a	N/a	N/a	N/a	N/a	Toyota's freedom of design concept
EU Capri Project (VW Estate)	1999	Fuel cell/ battery	15kW/ PEM	Ballard Mark 500 Series	N/a	N/a	N/a	N/a	Involved Johnson-Matthey, ECN, VW, and Volvo
HyMotion	2000	Fuel cell	75kW/ PEM	N/a	220mi 350km	N/a	86mph 140km/h	Methanol	13 gal. Of Liquid Hydrogen
HyPower	2002	Fuel cell/ super-capacitors hybrid	40kW/ PEM	Paul Scherrer Institute	94mi/ 150km	N/a	N/a	Compress. hydrogen	

VW

appendix

Fuel Cell Buses

Table created by Fuel Cells 2000
(available for downloading at www.fuelcells.org/info/charts/buses.pdf)

Bus Mfr.	Operation	Model	Year Shown	Engine Type	Fuel Cell Size/Type	Fuel Cell Mfr.	Range (mi/km)	Max. Speed	Fuel Type	Picture
Bus Manufacturing U.S.A., Inc.	Generation I of Georgetown University's program	30-foot Transit Bus	1994	Fuel cell/battery hybrid	50kW/Phosphoric Acid FC (PAFC)	Fuji Electric	250mi 402km	55mph 90km/h	Methanol	
Bus Manufacturing U.S.A., Inc.	Generation I of Georgetown University's program	30-foot Transit Bus	1995	Fuel cell/battery hybrid	50kW/PAFC	Fuji Electric	250mi 402km	55mph 90km/h	Methanol	
Bus Manufacturing U.S.A., Inc.	Generation I of Georgetown University's program	30-foot Transit Bus	1995	Fuel cell/battery hybrid	50kW/PAFC	Fuji Electric	250mi 402km	55mph 90km/h	Methanol	
NovaBus Corporation (a subsidiary of Volvo)	Generation II of Georgetown University's program. This bus will start a 1 yr. demonstration with Washington DC's Metro Area Transit Authority later in 2003	40-foot heavy duty transit buses	1998	Fuel cell/battery hybrid	100kW/PAFC Ambient-pressure	UTC Fuel Cells	350mi 563km	66mph 106km/h	Methanol	
NovaBus Corporation (a subsidiary of Volvo)	Generation II of Georgetown University's program. This bus is used for national demonstration purposes	40-foot heavy duty transit buses	2000	Fuel cell/battery hybrid	100kW/PEMFC	Ballard	350mi 563km	66mph 106km/h	Methanol	
Undetermined	Generation III of Georgetown University's program	40-foot low-floor bus platform	2003	Fuel cell/ultra capacitors hybrid	At least 240 kW/PEMFC	Undeter.	N/a	N/a	Methanol	
New Flyer Industries Ltd.	Proof of Concept	P1; low fl. transit bus based on New-Flyer model 40	1993 world's first	Fuel cell/battery hybrid	90kW/PEMFC	Ballard	250mi 400km	60mph 95km/h	Compress. Hydrogen	

	Proof of Concept		Year							
N/a		P2: full-sized, 40-foot	1995	Fuel cell/ battery hybrid	205kW/ PEMFC	Ballard	250mi 400km	N/a	Compress. Hydrogen	
Evobus: a Daimler Chrysler company	Accumulated over 540 hrs driving exper. By 1997; two week road test in Oslo, Germany 1999	**Nebus:** 405 low-fl. urban regular-service bus	1997	Fuel cell/ battery hybrid	205kW/ PEMFC	Ballard	155mi 250km	50mph 80km/h	Compress. Hydrogen	
Evobus: a Daimler Chrysler company	Demonstrated at SunLine Transit, AC Transit, and CaFCP	**Zebus (P4):** 40 ft. (1 year demo with SunLine)	1999	Fuel cell/ battery hybrid	205kW/ PEMFC	Ballard	N/a	N/a	Compress. Hydrogen	
Evobus: a Daimler Chrysler company	Sold as part of the CUTE; ECTOS; Perth, Australia programs. Cost ~US$3 million unsubsidized each. Madrid's 1st Citaro has been delivered.	**Citaro (P5):** (33 for the CUTE, ECTOS, STEP)	2003	Fuel cell/ battery hybrid	205kW/ PEMFC	Ballard	124mi 200km	50mph 80km/h	Compress. Hydrogen @ 5,000 psi	
Gillig Corporation	VTA, San Metro Transportation District, CaFCP & CARB - 3 FC Buses will be operated at VTA in San Jose, Ca	N/a	2004	Fuel cell/ battery hybrid	205 kW/ PEMFC	Ballard	N/a	N/a	Hydrogen	
Irisbus: a Renault V.I. and Iveco Co.	Will be demonstrated in Torino, Italy in 2002	40 foot	2001	Fuel cell/ battery hybrid	60kW/ PEMFC Ambient-pressure	UTC Fuel Cells	N/a	N/a	Compress. Hydrogen	
MAN "Bavaria 1"	Regular service in Erlangen and Nuremberg, Germany. 50% funded by Bavarian State	40 ft. low-floor city bus NL 263 "Bavaria I"	2000	Fuel cell/ battery hybrid	120kW/ PEMFC	Siemens	155mi 250km	50mph 80km/h	1548 L Compress. Hydrogen	

Manufacturer	Notes	Model	Year	Configuration	Power	FC maker	Range	Speed	Storage	Image
MAN	Will be used for EU's THERMIE program: Berlin, Copenhagen, Lisbon	40 ft. MAN N L223 low floor	Not Compl	Fuel cell/ Super capacitator hybrid	5 x 30kW/ PEMFC	Nuvera	N/a	N/a	700 L Liquid Hydrogen @ -253° C	
MAN	Will deliver one fuel cell bus to be operated as part of the hydrogen project at Munich Airport from early 2004 onwards.	40 ft. MAN low floor	Not Compl	Fuel cell/ battery hybrid	PEMFC	Ballard	N/a	N/a	H2 tanks on the roof at 5,000 psi	N/a
Neoplan	2 years fee-paying service in public traffic in the German spa resort Oberstdorf. Funded by Bavarian State	Midi bus N 8008 FC	1999	Fuel cell/ battery hybrid	40kW/ PEMFC	Nuvera	373mi 600km	30mph 50km/h	Compress. Hydrogen	
Neoplan	Available for Sales	N8012 - 33-seat bus	2000	Fuel cell/ 100kW flywheel hybrid	80kW/ PEMFC	Proton Motor Fuel Cell GmbH	155mi 250km	50mph 80km/h	Compress. Hydrogen	
New Flyer Industries	Demo. service of 3 buses in Chicago (1997) and Vancouver (1998) for 2 years	P3: H40LF models	1998	Fuel cell/ battery hybrid	205kW/ PEMFC	Ballard	N/a	N/a	Compress. Hydrogen	
New Flyer Industries	Natural Resources Canada (US$1.9 million) and Hydrogenics for demo in Winnipeg, Manitoba, Canada *Will incorp. Vehicle-To-Grid technology	40 ft.	March 2005	Distributed array of 25kW modules w/ ultra-capacitors	180kW/ PEMFC	Hydroge nics	N/a	N/a	Compress. Hydrogen	N/a
NovaBus Corporation (a subsidiary of Volvo)	Demonstrated in NY, NV, and DC. Received FTA funding to continue program.	Standard 40-foot transit bus	1999	Zinc-Air fuel cells with batteries	Zinc-Air	Arotech	N/a	65mph 105km/h	Zinc	
NovaBus Corporation (a subsidiary of Volvo)	Plans for RTC (Nevada Transit Agency) to use 2 – 5 buses	Standard 40-foot transit bus	2001	Zinc-Air fuel cells with ultra-capacitors	Zinc-Air	Arotech	N/a	N/a	Zinc	

Company	Notes	Model	Year	Type	Power/FC	Fuel Cell	Range	Speed	Fuel	Image
Thor Industries (ThuderPower LLC)	Will be tested by SunLine Transit in 2002 for 6 months (started public service at Sunline Nov. 6, 2002)	30 ft. Low Floor El Dorado National E-Z Rider	2001	Fuel cell/battery hybrid	75kW/ PEMFC Ambient-pressure	UTC Fuel Cells	200mi 322km	55mph 90km/h	Compress. Hydrogen	
Van Hool	3 will be used in regular service at AC Transit	40 foot	Not Compl	Fuel cell/battery hybrid	PEMFC Ambient-pressure	UTC Fuel Cells	250mi 400km	65mph 105km/h	5,000 psi Compress. Hydrogen	
NABI	1 will be used in regular service at SunLine Transit	45 foot	Not Compl	Fuel cell/battery hybrid	PEMFC Ambient-pressure	UTC Fuel Cells	N/a	N/a	Compress. Hydrogen	N/a
Van Hool	No Demonstration (Project EUREKA)	18 meter City Bus	1995	Fuel cell/battery hybrid	78kW/ PAFC	Elenco	186mi 300km	N/a	700 Liters Liquid Hydrogen	
Macchi-Ansaldo (EC project EQHHPP)	Company Testing only; part of the EC project EQHHPP	Full size regular floor city bus	1997	Fuel cell/battery hybrid	45kW/ PEMFC	Nuvera	250mi 400km	N/a	600 Liters Liquid Hydrogen	
Hino Motors Ltd. (Toyota subsidiary)	Toyota in-house testing	Low-floor city bus: FCHV-BUS1	2001	Fuel cell/battery hybrid	160kW/ PEMFC	Toyota	186mi 300km	50mph 80km/h	Compress. Hydrogen @ 5,000 psi	
Hino Motors Ltd. (Toyota subsidiary)	Tokyo metro. gov. will use this bus during summer 2003 on Tokyo's new waterfront route – Japan's nat'l debut of public fuel cell buses	60 pass. Low fl., diesel model: FCHV-BUS2	2002	Fuel cell/battery hybrid	180kW/ PEMFC [2 x 90kW]	Toyota	186mi 300km	50mph 80km/h	Compress. Hydrogen @ 5,000 psi	
NovaBus Corporation (a subsidiary of Volvo)	BVG - Berlin's public transportation body - to buy 2 prototypes, which will start testing in 2003	15.3 meter long Double-Decker	Not Compl	N/a	N/a	Proton Motor Fuel Cell GmbH	N/a	N/a	Hydrogen	Regular Bus Shown

appendix

Worldwide Hydrogen Fueling Stations

Table created by Fuel Cells 2000
(available for downloading at www.fuelcells.org/info/charts/h2fuelingstations.pdf)

Location	Fuel	Project	Dates	H2 Production Technique	Specifics/Comments	Picture
Auburn, California	Compress. H2	California Fuel Cell Partnership Station located at Pacific Gas & Electric service facility	Mid 2004	Ztek Corporation High Performance Steam Methane Reformer (HPSR)	HPSR converts natural gas supplied by PG&E to hydrogen at a rate of 600 scfh.	
Davis, California	Compressed H2, CNG/H2	University of California, Davis Hydrogen Bus Technology Validation Program, Toyota FCVs	In operation June 2003	Air Products delivered LH2	Converts liquid hydrogen to gaseous hydrogen.	
Riverside, California	Compress. H2	University of California, Riverside, College of Engineering – Center for Research and Technology with SCAQMD	1992 (1st of its kind)	Electrolyser Corp. (now Stuart Energy) Uni-polar electrolyzer capable of using PV array or grid operation for 5,000 psi H2.	"Solar-Hydrogen Production and Vehicle Refueling Station"	N/a
El Segundo, California	Compress. H2	Xerox Corp., DOE, UC Riverside, Matrix Engineers, City of West Hollywood, Kaiser Engineering, SCAQMD, CAN	Opened in 1995	Praxair fueling system; PVI Corp. photovoltaics; Stuart Energy hydrogen fueling station: electrolyzer	Electrolytic H2 generation "Clean Air Now Solar Hydrogen Vehicle Project"	N/a
Irvine, California	LH2 to Compressed H2	UC Irvine, SCAQMD, Toyota FCV	Opened June 2003	Air Products and Chemicals, Inc. will design and build this H2 fueling station	This station will support a fleet of Toyota's FCVs	
Thousand Palms, California	Compress. H2	SunLine Transit Agency and Ballard P4 Bus Demo.	Opened April 2000	Stuart Energy Hydrogen Energy Station for vehicle refueling	Electrolytic H2 generation and compression to 34.5 Mpa; 1,400 standard cubic feet per hour	
Sacramento, California	Liquid to Compressed H2, MeOH	California Fuel Cell Partnership BP, Shell, and Texaco helped in the design	Opened 11/2000	Air Products and Praxair delivered LH2	LH2 Stored on site in 4500-gallon tank. Can deliver CH2 to vehicle at 3600 and 5000 psi under 4 minutes. Uses Linde LH2 cryogenic nozzle and controls technology.	
Torrance, California	Compress. H2	American Honda Motors Co., Inc., Research and Development center	Opened 7/20/2001	Solar-powered electrolysis, stored on-site	PV-electrolysis with grid electricity back-up	

Location	Type	Organization	Date	Supplier/System	Description
Torrance, California	Compress. H2	As part of Toyota's efforts to establish California fuel cell "communities" with the leasing of 6 FCHVs to 2 UC campuses, it plans to open 5 more refueling stations in addition to this one.	Opened early 2003	Toyota is working with Stuart Energy and Air Products and Chemicals, Inc.	Toyota USA headquarters in Torrance uses a Stuart Energy hydrogen fueling station. It uses onsite electrolysis powered by renewable energy to generate 24kg hydrogen/day
Oxnard, California	Liquid H2	BMW North America Engineering and Emission Test Center	Opened 7/12/2001	Air Products delivered LH2	Manual power assisted refueling station. Also has a Linde LH2 mobile refueling station.
Chula Vista, California (mobile station)	Compress. H2	City of Chula Vista	2003	Stuart Energy hydrogen fueling station	A CFP-1350 generates 60 kg of H2/day, can fuel 3 buses a day, and dispenses at 3,600 and 5,000 psi.
Thousand Palms, California	Compress. H2	Schatz Hydrogen Generation Center at SunLine Transit	Opened 1994; retro fit in 2001-2	Teledyne Energy electrolyzer System	3600 psi hydrogen generation via electrolysis powered by renewable PV; produces up to 42 standard cubic feet per hour of H2
Richmond, California	Compress. H2	AC Transit facility	Opened 10/30/2002	Stuart Energy Satellite Hydrogen Energy Station for vehicle fueling	PEM electrolyzer; first satellite hub for CaFCP vehicles. Has 47 kg H2 storage capability.
San Jose, California	To Be Determined (TBD)	VTA, San Mateo Transportation District, CaFCP, and CARB	2004 readiness target	Air Products delivered LH2	Current fueling station at VTA's San Jose division will be enhanced with hydrogen capabilities
Chicago, Illinois	Liquid to Compress. H2 at station	Chicago Transit Authority – Ballard Bus Demo.	03/98 – 02/2000	Air Products delivered LH2	Three fuel cell buses
Crane, Indiana	Compress. H2	Navy Refueler	Delivered in 2004	Hydrogenics PEM Electrolyzer Refueler	20 kg/day of hydrogen

Location	Fuel	Project	Date	Supplier	Description	Image
Dearborn, Michigan	Liquid H2 & Liquid to Compress. H2 at Station	Ford Vehicle Refueling Station	Opened 1999	Air Products and Chemicals delivered hydrogen	N/a	
Ann Arbor, Michigan	LH2 to Compress. H2	EPA's National Vehicle and Fuel Emissions Laboratory (NVFEL), DaimlerChrysler, UPS	2004	Air Products and Chemicals Inc. will design and build this H2 fueling station.	This station will support a fleet of DaimlerChrysler FCVs used by UPS located at EPA's NVFEL. It will store up to 1,500 gallons of LH2	
Milford, Michigan	Compressed H2	GM and APCI	Will open 2004	GM and APCI	Integrates H2 generation, storage, and dual pressure (3,600 and 5,000 psi) dispensing all mounted on a single trailer.	N/a
Arizona (mobile station)	Compress. H2	Ford Motor Company	Delivered in 2002	Stuart Energy Mobile Hydrogen Energy Station for vehicle fueling	Integrates H2 generation, storage, and dual pressure (3,600 and 5,000 psi) dispensing all mounted on a single trailer.	
Phoenix, Arizona	Compress. H2, CNG, & H2/CNG blend	Arizona Public Service (Vehicle Testing Center – part of DOE Field Operations Program)	Opened in 2001	Proton Energy's HOGEN PEMFC electrolyzer	Only DOE / private sector H2 station	
Las Vegas	Compress. H2	Nevada Test Site Development Corp., DOE, Corporation for Solar Technologies and Renewable Resources and city of Las Vegas	Opened 11/15/02	Air Products and Chemicals	First multi-purpose station: H2 production via NG reformation; electricity production (for sale) using 50kW PEMFC; H2/CNG blends & pure H2 vehicle dispensers (uses Plug Power PEM fuel cell)	
Washington DC	LH2 & Compress. H2	General Motors Corp. and Shell Hydrogen	To be opened Oct. 2003 until 2005	Shell Hydrogen	US's 1st H2 pump at a Shell retail gas station. Will support a GM fleet of 6 H2 FCVs.	N/a
Penn State, Pennsylvania	Compress. H2	DOE, APCI, Penn State	Fall 2004	APCI, Onsite reformed natural gas	Commercial station project. Will H2 refuel up to 7000 psi.	N/a
Charlotte, North Carolina	Compress. H2	APCI and John Deere	2004	APCI and John Deere		N/a
Munich, Germany	Liquid H2	BMW Company Refueling Station	Opened in 1989	Linde	N/a	
Hamburg, Germany W.E.I.T. phase I	Compress. H2	W.E.I.T. hydrogen project Services hydrogen vehicles for: Hamburg Hermes Versand Service, HEW, and HHA	Opened on 12/01/99	Delivered Compressed H2 by m-tec Gastechnologie and Messer Griesheim	On-site electrolysis using 'Green' electricity and 100% fuel cell powered vehicles is the current goal/direction of this project	

Location	Type	Project / Partners	Opened	H2 Supply	Notes	
Hamburg, Germany W.E.I.T. phase II	Compress. H2	CUTE Bus Demo. PLANET from EUHYFIS in charge of H2 station	Opened 9/15/2003	Hamburgische Electricitätswerke AG subsidiary, GHW	On site hydrogen from electrolysis via renewable wind power. This is the second phase of the W.E.I.T. project, which will incorporate the Hamburg CUTE project	
Nabern, Germany	LH2 and LH2 to Compress. H2	DaimlerChrysler Company Refueling Station	Opened in 1998	LH2 delivered by Linde.	Uses Linde H2 refueling technology.	N/a
Munich, Germany	Compress. H2 & Liquid H2 & Liquid to Compress. H2	Munich Airport Vehicle Project Bavaria's minister for economics, transportation and technology and Linde	5/99	Hamburgische Electricitätswerke AG subsidiary, GHW LH2 and Compress. H2 delivered by Linde.	Uses Linde H2 refueling technology. Public Accessible.	
Wolfsburg, Germany	Liquid H2	On-site fueling for VW hydrogen vehicles	N/a	LH2 delivered by Linde.	Uses Linde refueling technology.	N/a
Russelsheim, Germany	Liquid H2 & Compress. H2	On-site fueling for Opel hydrogen vehicles	N/a	LH2 and Compress. H2 delivered by Linde.	Uses Linde LH2 and Compress. H2 refueling technology. This station can refuel Compress. H2 at 10,000 psi (700 bar).	N/a
Sindelfingen, Germany	Compress. H2 & Liquid H2	DaimlerChrysler	planned	H2 delivered by Linde.	Uses Linde H2 refueling technology. Can refuel to 5,000 psi.	N/a
Berlin, Germany	H2 & LH2 & Convent. fuels	Aral, BMW, BVG, DaimlerChrysler, Ford, GHW, Linde, MAN and Opel: Clean Energy Partnership (CEP)	2003 target	LH2 delivered by Linde to Aral station.	World's 1st public hydrogen gas station. Linde supplied all LH2 refueling technology.	
Berlin, Germany	Liquid and Compress. Hydrogen	TotalFinaElf, BVG, Linde, MAN and Opel: Hydrogen Competence Center Berlin. Station was opened under the framework of the Berlin, Copenhagen, Lisbon fuel cell bus Program	Opened 10/23/02	Uses Linde supplied LH2 & Proton Energy Systems' HOGEN® PEM electrolyzer for Compress. H2	1st permanent hydrogen fuel station in Berlin; will fuel H2 ICE buses from MAN & fuel cell buses Uses Linde LH2 refueling technology & includes a Linde AG mobile filling station	
Copenhagen	Mobile LH2	Station was opened under the framework of the Berlin, Copenhagen, Lisbon fuel cell bus Program	2002/3 target	Will use Linde supplied liquid hydrogen	The Linde mobile filling station is a part of the TotalFinaElf station in Berlin.	
Lisbon	Mobile LH2	Station was opened under the framework of the Berlin, Copenhagen, Lisbon fuel cell bus Program	2002/3 target	Will use Ariquido (in Portugal) supplied liquid hydrogen	The Linde mobile filling station is a part of the TotalFinaElf station in Berlin.	

Location	Type	Program	Date	Supplier	Production	
Erlangen, Germany	Mobile Liquid H2	MAN, Linde (several Bavarian funded bus programs)	4/12/96 – 8/98 (ICE) & again in 10/2000 – 04/2001 (fuel cell)	Linde AG produced and supplied the LH2 to their mobile station	Linde AG supplied LH2 from their large central H2 production & Liquification plant and transported it to the Linde mobile fueling station	
Oberstdorf Spa, Germany	Compress. H2	Neoplan fuel cell bus at Oberstdorf; funded by Bavarian State	1999 – 2001	Linde AG produced and supplied the LH2 to their mobile station	Linde AG supplied LH2 from their large central H2 production & Liquification plant and transported it to the Linde mobile fueling station	N/a
Stuttgart, Germany	Compress. H2	CUTE Bus Demo.	2003 target	BP affiliated	On-site Natural Gas steam reformation	N/a
Malmo, Sweden	Compress. H2	Sydkratt	Opened 9/11/2003	Stuart Energy's Hydrogen Energy Station for vehicle fueling	Water electrolysis-based hydrogen generation	
Stockholm, Sweden	Compress. H2	Clean Urban Transport for Europe (CUTE) Bus Demo. PLANET from EUHYFIS in charge of H2 station	11/13 2003	Stuart Energy's Hydrogen Energy Station for vehicle fueling	Central Hydro Powered electrolysis, then transported to fueling site	N/a
London, United Kingdom	Compress. H2	CUTE Bus Demo. BP in charge of H2 station	2003 target	BP affiliated	Centralized production via excess hydrogen from crude oil, then transported to fueling site. Uses BOC H2 refueling technology.	
Amsterdam, The Netherlands	Compress. H2	CUTE Bus Demo. GVB Amsterdam in charge of H2 station	2003 target	Hydrogen System's IMET® powered water electrolyzer and Hoekloos (a Linde Co.) delivered Compress. H2	On site Hydrogen production via electrolysis from green energy	N/a
City of Luxemburg	Compress. H2	CUTE Bus Demo. AVL in charge of H2 station	Opened 10/13/2003	Shell Hydrogen/Air Liquide delivered	Tube Trailer compressed H2	N/a
Porto, Portugal	Compress. H2	CUTE Bus Demo. BP in charge of H2 station	Late 2003	BP affiliate	Centralized production via excess hydrogen from crude oil. Uses a Linde High Booster Compressor System for high pressure H2.	N/a
Madrid, Spain	Compress. H2	CUTE Bus Demo. & CITYCELL ETA in charge of H2 station	Opened 4/28/2003	Steam reforming of natural gas	Consortium (AIR LIQUIDE, gasNatural, REPSOL YPF)	

Location	Type	Project / Sponsor	Status	Production	Description	
Barcelona, Spain	Compress. H2	CUTE Bus Demo. BP in charge of H2 station	Opened 9/22/2003	BP & Vandenborre Hydrogen Systems: IMET® powered water electrolyzer	On-site production via renewable solar and grid electricity powered electrolysis. Uses a Linde High Booster Compressor® System for high pressure H2.	
Europe	Compress. H2	EU, Bauer Kompressoren, Casale Chemicals, PLANET (EUHYFIS Project)	R&D phase complet. 1st demo station in 2004	N/a	On-site electrolysis of water from a renewable electrical source (solar or wind). Currently contributing to the CUTE program	
Reykjavik, Iceland	Compress. H2	ECTOS Bus Demo.	Opened April 2003	Shell Hydrogen/Iceland	World's 1st Commercial Hydrogen Station. On site Geothermal and Hydro Powered Electrolyzer.	
Perth, Australia	Compress. H2	DaimlerChrysler, BP, UNEP	2004 target	Centrally produced H2 at BP's refinery in Kwinana	Centralized hydrogen production. Gaseous hydrogen from crude oil/natural gas refining.	N/a
Victoria, Australia	Compress. H2	One H2 fueling station to service several hydrogen fuel cell buses taking passengers to and from the Victorian Fast Train (program is under review)	TBD	TBD	Reviewing electrolysis via solar and reforming natural gas	N/a
Beijing, China	To be determined (TBD)	Global Environment Facility (GEF) and United Nations Development Program (UNDP): commercial demonstrations of 6 fuel cell buses and hydrogen refueling stations.	TBD	N/a	N/a	N/a
Osaka, Japan	Compress. H2	PEMFC Vehicle Demo. by WE-NET	Fall 2001 – end of 2003	N/a	Natural Gas Reforming	
Takamatsu, Japan	Compress. H2	PEMFC Vehicle Demo. by WE-NET	Fall 2001 – end of 2003	N/a	PEM electrolyzer	
Tsurumi, Japan	Compress. H2	PEMFC Vehicle Demo. by WE-NET	Opened 8/2002	N/a	N/a	N/a
Yokohama, Japan	Compress. H2	Cosmo Oil JHFC	Opened FY2002	N/a	Part of Japan Hydrogen and Fuel Cell Demonstration Project, which will build 5 H2 stations in Tokyo	N/a

Location	Type	Owner / Partners	Date	Technology	Notes	
Yokohama, Japan	Compress. H2	Nippon Oil JHFC	Opened FY2002	N/a	Desulfurized-gasoline Reformation. Part of Japan Hydrogen and Fuel Cell Demonstration Project, which will build 5 H2 stations in Tokyo. Naphtha Reformation	
Japan	Compress. H2	Company Filling Stations for Honda	Opened in 2001	N/a	N/a	N/a
Japan	Compress. H2	Company Filling Station at Toyota	Opened in 2001	N/a	N/a	N/a
Tokai, Japan	Compress. H2	Toho Gas Co. owned. Will sell the hydrogen fuel at a price similar to gasoline	Opened 10/2002	N/a	Located at Toho Gas Co.'s research laboratory in Aichi Prefecture	N/a
Tokyo, Japan	Liquid H2 & Compress. H2	Iwatani International Corporation; Tokyo Metropolitan Government; Showa Shell Sekiyu KK JHFC	Target: April 2003 2 year operation	LH2 from Iwatani and high pressure Compress. H2 from Linde Hydrogen Cryo-Compressor.	Tokyo's first hydrogen station. Uses Linde LH2 refueling technology.	
Kawasaki City, Japan	Compress. H2	Air Liquide Japan JHFC	2003 target	N/a	Methanol Reformation	
Tokyo, Japan	Compress. H2	Tokyo Gas & Nippon Sanso JHFC	N/a	Senju	LPG reforming	
Toronto, Canada	Compress. H2	Hydrogenics, City of Toronto, and the Canadian Transportation Fuel Cell Alliance	2003	Steam Methane Reforming	70 kg/day of hydrogen; this hydrogen energy station includes a 50kW PEM fuel cell for peak-shaving.	
Toronto, Canada	Compress. H2	Hydrogenics, City of Toronto, and the Canadian Transportation Fuel Cell Alliance	2004	Hydrogenics PEM Electrolyzer Refueler	65 kg/day of hydrogen; electricity renewable resources	
Toronto, Canada	Compress. H2	Hydrogenics, John Deere, NACCO, FedEx, GM of Canada – Fuel Cell forklift demonstration	2004	Hydrogenics PEM Electrolyzer Refueler	20 kg/day of hydrogen	N/a

Location	Type	Project	Date	Technology/Company	Notes	Image
Toronto, Canada	Compress. H2	Hydrogenics, Purolator, and and the Canadian Transportation Fuel Cell Alliance – Fuel Cell Hybrid Delivery Truck Demonstration	Late 2004	Hydrogenics PEM Electrolyzer Refueler	20 kg/day of hydrogen	N/a
Vancouver, Canada	Compress. H2 & H2/Natural Gas blend	British Columbia Hydro's Powertech Labs	Opened in 2001	Stuart Energy hydrogen fueling station: electrolyzer	Used for Coast Mountain Transit's fuel cell bus demonstration from '98-00. It now supplies H2 as well as a blend of H2/Natural Gas to a variety of vehicles.	N/a
Surrey, BC Canada	Compress. H2	BC HydroGen	Opened Fall of 2001	N/a	70 Mpa hydrogen via electrolysis from renewable energy	N/a
Torino, northern Italy	Compress. H2	Irisbus PEMFC City Bus Demo.	2003 target	N/a	Hydrogen from hydropower via electrolysis	N/a
Bi-cocca (near Milano)	Compress. H2 & Liquid H2	Hydrogen and fuel cell demonstration project	Opened in 2002	AEM, SOL, and others	Hydrogen liquefier and vehicle refueling	N/a
Oostmalle, Belgium	Liquid H2	Belgian Bus Demo.	Opened in 1994	Messer Griesheim GmbH	LH₂ storage system of 125 L, an electric LH₂ evaporation system as well all necessary connecting supply infrastructure and relevant control and safety components	N/a
Leuven, Belgium	Compress. H2	NexBen Fueling—a division of Chart—has won a contract from Citensy	2003	NexBen Fueling	Europe's first combined liquefied natural gas (LNG) and liquid compressed natural gas (LCNG) and hydrogen fueling station	N/a
South Korea	Compress. H2	Hyundai Motor Company fuel cell vehicle research	Opened in 2001	Pressure Products Industries, Inc. & Doojin Corporation	The heart of the fueling station is a PPI two stage compressor, model 4V104068 designed for 6,000 psig	N/a
Singapore	Compress. H2	Singapore Economic Dev. Board (EDB), SINERGY Programme (Singapore initiative in energy technology), BP, DaimlerChrysler, APCI	1st quarter 2004	Air Products	This will be the 1st H2 refueling station in Southeast Asia. Will be able to supply 20kg of compress. H2 per day (~11 vehicles per day). This will support a small fleet of DaimlerChrysler F-Cell FCVs	N/a
Taoyuan, Taiwan	Compress. H2	Maw Chong Energy Hydrogen refueling project	2004	Ztek Corporation High Performance Steam Methane Reformer (HPSR)	Ztek to deliver 2000 scfh unit in 2004	N/a
Howaldwerke-Deutsche Werft, Germany		Class 212 submarine: driven by hydrogen fuel cells dependent on outer air.	Finished in 2002	Air Products (USA)	World's 1st installed complete hydrogen infrastructure in a non-nuclear hydrogen driven submarine.	N/a

appendix

Worldwide Fuel Cell Installations

Table created by Fuel Cells 2000
(available for downloading at www.fuelcells.org/info/charts/FCInstallationChart.pdf)

Fuel cell Manufacturer	Project Partners	Fuel cell	Location	Building	Start date	Status	Fuel used	Picture	Comments / Contact Information
American Fuel Cell Corp., Plug Power, H Power, DCH Enable	**EPRIGEN** Buy Down Recipient FY1998	Sixteen 3 kW PEM systems	Palo Alto, CA	Testing Facility	1999	Compl.			EPRIGEN performs many tests on fuel cell units they've purchased or they go on-site to the manufacturer.
Multiple US and foreign-based FC manufacturers including Acumentrics, Avista Labs, Ballard, Enable, HyRadix and Plug Power	**HARC** with Dana Corp., Texaco, Southern Co., Salt River Project Sponsorship open to all organizations except fuel cell manufacturers	Various PEM and SOFC models and sizes intended for residential, UPS and APU markets Avista Labs' 3kW is First demo	The Woodlands, Texas	Houston Advanced Research Center (HARC)	1999	Ongoing	Natural Gas, Hydrogen and Propane		HARC's program is to evaluate fuel cell equipment and verify manufacturer performance claims. HARC: Dan Bullock dbullock@harc.edu 281-364-6087 HARC: (281) 363-7908 Barbara Peyton bpeyton@harc.edu
Ansaldo Fuel Cells Spa (AFCO)		100 kW MCFC	ENEL site near Milan, Italy		98-99	Compl.	Natural Gas		Proof of Concept
Ansaldo Fuel Cells Spa (AFCO)		100 kW MCFC	IBERDROLA site of Guadalix, near Madrid (Spain)	stack test and conditioning facility	1999	Compl.	Natural Gas		

Apollo Energy Systems Inc.	Apollo Energy and Hydrolec Inc.	10 kW Apollo Alkaline fuel cell w/ 12 kWh lead-cobalt battery	Ft. Lauderdale, FL		2002	In production			2,000 Apollo power plants per month to Hydrolec used for power back up and elevator systems around the world; starting in 2002. Worth $223.0 million. Recently acquired Zetek's AFC factory.
Ballard	Ballard / Belgium	250 kW Stationary Generator	Liege, Belgium		2001	On going			
Ballard	Alstom / Bewag AG	250 kW Stationary Generator	Berlin, Germany	Bewag's Treptow Heating plant	June 2000	On going	Natural Gas		The BEWAG project is for public education. It involves a closed fuel circuit comprised of photovoltaics and an electrolyzer.
Ballard	Cinergy Technology Inc.	250 kW Stationary Generator	Crane, Indiana	Crane Naval Surface Warfare Center	Sept. 1999	Ended 2001			First 250 kW PEM fuel cell generator in the world to enter field testing. Each Ballard PEM power plant will undergo 2 years of testing with the overall program complete in 2004
Ballard	Nippon Telegraph & Telephone (NTT) / Ebara	250 kW Stationary Generator	Tokyo, Japan	NTT research lab	2000	On going			This unit utilizes a cogeneration system incorporating an adsorption chiller (for air conditioning) developed by EBARA.
Ballard	Ballard / Japan	250 kW Stationary Generator	Tokyo, Japan		2001	On going			

	Project	System	Location	Site	Year	Status	Fuel		Notes / Contact
Ballard	Alstom / Elektra Birseck (EBM)	250 kW Stationary Generator	Basil, Switzerland	EBM headquarters	2000	On going	Natural Gas		The overall energy concept contains a PEMFC (polymer electrolyte membrane fuel cell) as a container plant for the generation of electricity and heat in combination with a microturbine http://www.pem-oberhausen.de/englisch/index.html
Ballard	"PEM-Oberhausen" project	250 kW Stationary Generator CHP with Microturbine	Fraunhofer Institute for Environmental, Safety and Energy Engineering (UMSICHT)				Natural Gas		
DCH Tech.	DOD Residential Fuel Cell Demonstration Program Brooks AFB	3 kW (3 units)	San Antonio, TX	Base Housing	Oct. 2002	Not installed	Natural gas		Company no longer in business.
DCH Tech.	DOD Residential Fuel Cell Demonstration Program Ft. Bragg	3 kW (1 unit)	Fort Bragg, NC	TBA	Dec. 2002	Not installed	Natural Gas		Company no longer in business.
Fuel Cell Technologies (FCT)	The Presidio Trust Buy Down Recipient FY2000	One 5 kW SOFC system	San Francisco, CA		Second half of 2002				Gary Allen, Marketing, [ext 114] gallen@fct.ca Phone: (613) 544-8222
Fuel Cell Technologies (FCT)	NFCRC	One 27 kW SOFC system	University of California, Irvine	Demonstration unit	Since 1998; new 5 kW unit in 2002	Used as a teaching model	Various – NG/Diesel /JP-8		2 generations of Siemens fuel cells: 1) white tubes and 2) black tubes Gary Allen, Marketing, [ext 114] gallen@fct.ca Phone: (613) 544-8222
Fuel Cell Technologies (FCT)	US National Park Service, BPA	Two 5 kW SOFC CHP systems	Yosemite National Park, California	Administration building	Never Installed		Propane		Two Plug Power systems will be installed early 2004 Mira Vowles 503.230.4796 mkvowles@bpa.gov Gary Allen: gallen@fct.ca

Company	Project	Size	Location	Facility	Date	Status	Fuel	Image	Contact
Fuel Cell Technologies (FCT)	Hammarby Sjostad project	Two 5 kW SOFC CHP systems	Stockholm, Sweden		August, 2002	To be delivered			
Fuel Cell Technologies (FCT)	AQMD	One 5 kW SOFC CHP systems	California	AQMD	End of 2002	To be delivered			
FuelCell Energy with Yale	Connecticut Clean Energy Fund and Yale University	250 kW	New Haven, CT	Environmental Science Center	Dec. 2003	Dedicated	Natural Gas		
FuelCell Energy	PPL Energy Plus, Starwood Hotels and Resorts	250kW direct fuel cell	New York, NY	Sheraton New York Hotel and Towers	April 2004	To be installed			
FuelCell Energy	MTU	250 kW DFC (Hot Module #1)	Bielefeld, Germany	University of Bielefeld	11/99		Natural Gas		
FuelCell Energy	Salt River Project	250kW Direct fuel cell	Mesa, Arizona	Arizona State University East Campus	Late 2004, early 2005	To be delivered			
FuelCell Energy	LOGANEnergy, US Coast Guard Station	5kW PEM Fuel Cell	New Orleans, LA	US Coast Gurard Station	November 2003	In operation			Sam Logan LOGANENERGY 770-650-6388 samlogan@loganenergy.com George Dunn US Coast Guard Station New Orleans, LA 504-846-6179 gdunn@staneworleans.uscg.mil

Company	Project	Size	Location	Application	Date	Status	Fuel	Image	Notes
FuelCell Energy	USCG Air Station Cape Cod Fuel Cell Project	250 kW MCFC	Cape Cod, MA	USCG Air Station (w/ hot water)	May 2003	Installed and Tested	Natural Gas		Involves: PPL Spectrum, PPL Savage Alert, Millennium Builders, Fuel Cell Energy, Mass Tech Collaborative, NSTAR, and the Coast Guard (ASCC, CEU Providence, R&D Center)
FuelCell Energy	City of LA / American Airlines Air Freight, Buy Down Recipient FY2000	One 250 kW MCFC system	Los Angeles, CA	American Airlines Freight					
FuelCell Energy	Ppl Spectrum, Inc. Buy Down Recipient FY2000	One 250 kW MCFC system							
FuelCell Energy with MTU		250 kW DFC (Hot Module #2)	Bad Neustadt, Germany	Rhoen-Klinikum AG Hospital	10/00				
FuelCell Energy	RWE AG, MTU Daimler Chrysler AG	250 kw	Essen Germany		July 2003	Joint venture established			
FuelCell Energy	MTU CFC Solutions Gmbh		Karlsruhe, Germany	Michelin tire plant	Feb 2003	Begun operation			
FuelCell Energy	Wabash River Energy, IGCC, Buy Down Program	2 MW	Wabash River Energy, IGCC	Clean coal operating plant	10/2003	Final stages of construction	combination of coal and renewable fuels		It is the first Clean Coal Technology plant to employ a fuel cell and run on a combination of coal and renewable fuels. Originally set to operate at the still under development Kentucky Pioneer Energy IGCC site, it has been moved to Wabash River in order to start operation 2 years ahead of schedule.

Company	Project	Size/Type	Location	Facility/Site	Date	Status	Fuel	Image	Contact/Notes
FuelCell Energy	FuelCell Energy Facility Demonstration	250 kW Direct FuelCell	Danbury, CT	FuelCell Energy's Facility	Feb. 1999	Off line as of June 30, 2000			It was grid-connected. After post-demo analysis, this plant will be modified to accommodate both a fuel cell and a gas turbine for another demonstration.
FuelCell Energy	Alabama Direct FuelCell Demonstration Project (Hot Module)	250 kW DFC	Tuscaloosa, AL	Mercedes Benz Production Facility	Dec 2003	Completed			7260 Btu/kWh = rated output
FuelCell Energy	DOE Harrison Mining Corp	200kW DFC	Hopedale, OH	Coalmine Methane Test Site			Coalmine Methane Gas		Steven Eschbach Director, Investor Relations & Communications 203-825-6000 seschbach@fce.com www.fuelcellenergy.com
FuelCell Energy	King County Digester/Fuel Cell Demonstration	1 MW DFC	King County, Washington	Municipal waste treatment system	2003	Under construction	Methane rich gas (digester gas)		http://dnr.metrokc.gov/wtd/fuelcell/ Greg Bush, Program Manager King County Wastewater Treatment Division 201 S. Jackson St., MS KSC-NR-0512 Seattle, WA 98104-3855 Tel: 206-684-1164 Fax: 206-684-2057

Manufacturer	Customer / Program	Type	Location	Site	Operation started	Status	Fuel	Image	Contact
FuelCell Energy	Los Angeles Department of Water and Power / US Army Coprs pf Engineers / Buy Down Program Recipient FY1999	250 kW molten carbonate DFC 300	Downtown LA	LA Depart. of Water and Power headquarters	Dec 2003	Trial plant replaced with permanent model	Natural Gas		Kjell Ostensen Kjell.ostensen@ladwp.cp; 213.276.0011
FuelCell Energy	Los Angeles Department of Water and Power / Buy Down Program Recipient FY2000	250 kW molten carbonate	Los Angeles, CA	Sewage plant	Planned for later in 2002	To be delivered in 2002	Digester gas from sewage plant		Bill Glauz, Manager of Distributed Generation Los Angeles Department of Water and Power; Phone = 213 267-0410; Fax = 213 367-0777; william.glauz@ladwp.com
FuelCell Energy	Los Angeles Department of Water and Power / Buy Down Program Recipient FY2001	250 kW Molten Carbonate	Los Angeles, CA	LA International airport	Planned for later in 2002	To be delivered in 2002			Bill Glauz, Manager of Distributed Generation Los Angeles Department of Water and Power; Phone = 213 267-0410; Fax = 213 367-0777; william.glauz@ladwp.com
FuelCell Energy	Los Angeles Department of Water and Power (LADWP)	2 MW DFC power plant	Santa Clara, CA	Scott Receiving Station	mid-1990s	Completed			First full scale utility demonstration of a molten carbonate fuel cell system
FuelCell Energy	Los Angeles Departmnet of Water and Power and Los Angeles Deprtment of Public Works and Sanitation / US Army Corps of Engineers / Buy Down Program	250kW Direct Fuel Cell	Terminal Island, San Perdo, CA		Operation started	May 2003	Natural Gas		Kjell Ostensen kjell.ostensen@ladwp.com 213.367.0011 Plans to operate on sewage gas by summer 2004

Company	Customer	Capacity	Location	Project	Date	Status	Fuel		Contact / Notes
FuelCell Energy	Starwood Hotels	250 kW	Parsippany and Edison, NJ	2 Starwood Sheraton Hotels	Fall 2003	Installation and Testing complete	Hydrogen		Partnered with New Jersey Clean Energy Program ($1.7 million).
FuelCell Energy	Grand Valley State University, Michigan Public Service Commission		Muskegon, MI	Research Center	May 2003	Under Construction			
FuelCell Energy	USDOE, Global Energy Inc	2 Megawatt Direct Fuel cell	Terre Haute, IN	Wabash River Energy, Ltd. Integrated Gasification Combined Cycle (IGCC) Facility	Fourth Quarter 2003	Operational	Natural Gas		Steve Eschbach, CFA, FuelCell Energy, Inc., +1-203-825-6000, seschbach@fce.com
FuelCell Energy	Zoot Enterprises	250kW	Brozeman, MT	Galactic Park High Technology Campus	January 2004	Installed	Natural Gas		Bruce Nelson Zoot Enterprises, Inc. 555 Zoot Enterprises Lane Bozeman, MT 59718 Phone: (406) 586-5050; Fax: (406) 586-8005; E-mail: bnelson@zootweb.com DOE Technology Development Manager: Chris Bordeaux Phone: (202) 586-3070; Fax: (202) 586-9811; E-mail: Christopher.Bordeaux@ee.doe.gov

Company	Customer / Partner	Size	Location	Site	Date	Status	Fuel	Notes
FuelCell Energy	US Coast Guard Research and Development Center		Virginia Beach, VA	Cape Henry Lighthouse	Installed 3/2002	Ongoing Evaluation after 6 months	Methanol and water mixture	US Coast Guard is servicing and maintaining this system. Said to use only 16 gallons of fuel / day whereas a generator uses 16 gallons of fuel / hour!
FuelCell Energy	Caterpillar Inc	250kW Direct Fuel Cell	Peoria, IL	Caterpillar Techinical Center	December 2003	Operational		For FuelCell Energy: Steve Eschbach, CFA 203-825-6000 seschbach@fce.com For Caterpillar: Carl M. Volz 309-675-5819 volz_carl_m@cat.com
FuelCell Energy	Siemens Westinghouse and BP	250kW	Nikiski, AK	BP's gas-to-liquid test facility	Aug 2001	To be installed 2003	Natural Gas	
FuelCell Energy with Marubeni Corporation	Kirin Brewery	250 kW DFC	Outside Tokyo, Japan	Industrial wastewater facility at Kirin Brewery	January 2003	operational	Digester Gas	
FuelCell Energy with Marubeni Corporation	Marubeni Corporation (FuelCell Energy's Asian Partner)	One 250 KW in 2001, then 4 additional 250 kW or a MW-class plant	New York, NY and Tokyo, Japan					Total contract is for 1.25 MWs and $6.25 million Takeo Nakata: Nakata-t@jp.marubeni.com

Company	Partner/Customer	Size/Type	Location	Facility	1st quarter 2003	2-year demo. program	Fuel	Image	Notes
FuelCell Energy with Marubeni Corporation	Municipality of Fukuoka, Japan	250 kW DFC	Fukuoka, Japan	Municipal wastewater treatment facility	1st quarter 2003	2-year demo. program	Methane-like digester gas		
FuelCell Energy	State University of New York College of Environmental Science and Forestry (SUNY-ESF), NYSERDA, NYPA	250 kW MCFC CHP	New York, New York	SUNY-ESF	2002	To be installed early 2003	Natural Gas		Waste heat for campus hot water, space heating and/or cooling. Pursued by New York Governor George Pataki. $1 million grant from NYSERDA
FuelCell Energy	PPL Energy Services (FuelCell Energy partner) New Jersey Clean Energy Fund	259 kW MCFC CHP	Toms River, New Jersey	Ocean County College	2003	To be Installed	Natural Gas		
Fuji Electric		Two 100 kW PAFC	Yamagata city, Japan	Sewage Treatment Center	6/2002	Operational;	Methane Gas		Fuji Electric has been developing PAFC technology since the 1980s. 50 kW and 100 kW PAFC systems have been commercially available since the late 1990s. They have almost sold 100 units thus far.
Fuji Electric	Tokyo Electric Power Co. (TEPCO)	50 kW	Japan	TEPCO New Energy Park	June 14 1993	Completed Dec. 1996			This unit accumulated 39,291 hours by 10/00.
Global Thermoelectric	Global Thermoelectric	1.35 kW			2000	Completed in June 2000	Natural gas		Operated over 1100 hours

Company	Program	Size/Type	Location	Site	Date	Status	Fuel	Image	Contact
Global Thermoelectric	Global Thermoelectric and Enbridge Inc.	Prototype testing	Enbridge (Canada)		2001	On Going			
Global Thermoelectric	Bonneville Power Administration	Three SOFC systems (2 to 5 kW)	BPA	Field testing sites	2003	To be delivered	Natural gas or propane		
GM	Prototype testing	75 kW PEMFC	Rochester, NY	GM's new fuel cell research facility	2001	On Going	Natural gas, methanol, gasoline		DOW Chemical Harold Nicoll 989-636-5162 hgnicoll@dow.com
GM	DOW Chemical	75kW	Freeport, TX	DOW Chemical site	Feb 2004	Operational for next 4-6 months			GM Scott Fosgard 586-947-3295 scott.fosgard@gm.com
H Power	DOD Residential Fuel Cell Demonstration Program Sierra Army Depot	4.5 kW (1 unit)	Herlong, CA	Barracks	Sept. 2002	Not installed	Propane		Company bought out by Plug Power
H Power	DOD Residential Fuel Cell Demonstration Program Patuxent River NAS	4.5 kW unit	Patuxent River, MD	Office building	Oct. 2002	Not installed	Propane		Company bought out by Plug Power
H Power	DOD Residential Fuel Cell Demonstration Program Patuxent River NAS	4.5 kW unit	Patuxent River, MD	Base Housing	Oct. 2002	Not installed	Natural gas		Company bought out by Plug Power

Manufacturer	Customer / Partner	System	Location	Application	Date	Status	Fuel		Notes
H Power	Kamata Inc. (Leading Japan propane distributor)	4.5 kW cogeneration system	Gotenba-city, Japan	Kamata's Gotenba employee facility	May 2002	Installed	Propane		Marks first field test of a propane fueled PEM system for an actual load in the Japanese residential market.
H Power	Gaz de France	Six 4.5 kW beta cogeneration units	Gaz de France's research and development campus in La Plaine Saint Denis, a suburb of Paris	Gaz de France model home	2001	Being installed	Natural Gas		
H Power	Naps Systems Oy (Naps) in Finland, Birka Energi, and ABB	4 kW cogeneration system	Hammarby Sjostad Stockholm, Sweden	Environmental Information Center	June 4, 2002	Installed	H_2 produced via PV system comb'ed with an electrolyzer		A second source of hydrogen will be provided from biogas produced at a nearby municipal waste plant
H Power	Energy Co-Opportunity (ECO) – H Power's marketing partner	50 EPAC™- 500 Watt units	Used by various member cooperatives of ECO	Outdoor applications	Order placed 5/2002	Delivered	Direct Hydrogen		Telecom. towers, radio transmitters, signal lights, billboards, emergency back-up for critical circuits
H Power	Energy Co-Opportunity (ECO)	Four 4.5 kW co-generation residential units			8/2002	Installed			Part of the beta test program with partner Energy Co-Opportunity.
H Power	Fall River Rural Electrical Cooperative of Ashton, Idaho (member of ECO) and The Propane Education & Research Council	4.5 kW cogen. system	Yellowstone National Park, Wyoming	Ticket kiosks and an office at the West Entrance at the West Yellowstone	4/2002	Installed	Propane		Company bought out by Plug Power

Company	Partner	System	Location	Site	Shipped	Status	Fuel	Image	Notes
H Power	Mitsui & Co., Ltd. (H Power's Asian marketing partner) and Osaka Gas	Eight 500 Watt Beta PEMFC residential cogeneration units	Kansai area of Japan	In-house at Osaka Gas and field locations	Shipped by 2/2003		Natural Gas		Uses Osaka Gas's proprietary steam reforming technology
H Power	Mitsui & Co., Ltd. (H Power's Asian marketing partner) and Osaka Gas	Alpha PEMFC residential cogeneration units	Kansai area of Japan	In-house at Osaka Gas and field locations	1/2002	Testing moved to Beta units	Natural Gas		Uses Osaka Gas's proprietary steam reforming technology
H Power	HavePower	1 EPAC™ back up system	Outside Wilkes Barre, PA	Cellular Comm. Tower	2/21/2002	On Going			World's first cellular communication's tower powered by hydrogen fuel cells.
Hydrogenics, Corp.	Nextel	25 kWHyUPS regen. back-up	Northern CA	Nextel Comm. Trial site	10/2001	9/2003			Target market for this product is the demanding telecommunications industry.
HyRadix Inc	DOE, Propane Education and Research Council, Texas Fuel Cell Partnership		San Antonio, Texas	Texas DOT TransGuide Facility	4/2004	In operation	Propane		First propane fuel cell Technical Contact: Dan Kelly dan.kelly@rrc.state.tx.us Partnership Contact: Ken Zarker kzarker@tceq.state.tx.us
IdaTech	RWE Fuel Cells	2.5kW	Berlin, Germany	Representative office of the State of the North Rhine-Westphalia	4/2004	To be installed	Natural gas		

Company	Program	System	Location	Customer	Date	Status	Image	Notes
IdaTech	IdaTech and BPA Alpha unit Testing Buy Down Program Recipient FY1999	5 kW PEM Alpha system	Redmond, OR	Central Electric Co-Operative	Winter 2003	No longer operational		First commercial testing of residential fuel cell systems in NW. 6 were installed at participating utilities for the initial phase of the test (Alpha) fuelcells@bpa.gov Mira Vowles 503.230.4796 mkvowles@bpa.gov
IdaTech	IdaTech and BPA Alpha unit Testing	5 kW PEM Alpha system	Eugene, OR	Emerald Peoples Utility Distribution	Winter 2003	No longer operational		Mira Vowles 503.230.4796 mkvowles@bpa.gov
IdaTech	IdaTech and BPA Alpha unit Testing	5 kW PEM Alpha system	Rotating location	PNGC	Winter 2003	No longer operational		Mira Vowles 503.230.4796 mkvowles@bpa.gov
IdaTech	IdaTech and BPA Alpha unit Testing	5 kW PEM Alpha system	Richmond, WA	Energy Northwest Mobile	Winter 2003	No longer operational		Mira Vowles 503.230.4796 mkvowles@bpa.gov
IdaTech	IdaTech and BPA Alpha unit Testing	5 kW PEM Alpha system	Eureka, MT	Lincoln Electric	Winter 2003	No longer operational		Mira Vowles 503.230.4796 mkvowles@bpa.gov
IdaTech	IdaTech and BPA Alpha unit Testing	5 kW PEM Alpha system	Fergus Electric	Lewiston, MT	Winter 2003	No longer operational		Mira Vowles 503.230.4796 mkvowles@bpa.gov
IdaTech	IdaTech and BPA Beta unit Testing	5 kW PEM Beta system	Various - undisclosed		Winter 2003	No longer operational		Mira Vowles 503.230.4796 mkvowles@bpa.gov
IdaTech	Research & Development division of Electricite de France (EDF)	5 kW PEM system	France	R&D division of (EDF)	2/01	Will end Spring 2002		This field test helps lay the groundwork for IdaTech's commercialization of portable and residential systems. fuelcells@bpa.gov

Company	Partner	System	Location	Site	Date	Status	Fuel	Notes
IdaTech	Portland General Electric through a Bonneville Power Administration program	5 kW PEM system	Portland, Oregon	Portland General Electric Earth Advantage National Center	Jan 2004	Operational		Mira Vowles 503.230.4796 mkvowles@bpa.gov Part of BPA's Northwest demonstration program First fuel cell in Oregon connected to a power grid
Industrial Research Laboratory (IRL)	Australian Cooperative Research Center for Renewable Energy	6 kW	Perth, Australia	Wind turbine	2003		Hydrogen	The system will link to a wind turbine powering an electrolyser to generate the hydrogen. The fuel cell will serve as back up when the wind turbine is not capable of producing power.
Matsushita	Japan Gas Association	PEM CHP residential cogeneration systems	Japan	In-house testing at Japan Gas Assoc.	2001	On going	Natural gas	
Mosaic Energy	NiSource Inc., Ishikawajima-Harima Heavy Industries, and the Gas Technology Institute	5 kW PEMFC	Yokohama, Japan	Retail gasoline service station	July 26, 2001	Testing into 2002	Naphtha	World's first liquid fuel PEMFC test
Mosaic Energy	NiSource Inc., Ishikawajima-Harima Heavy Industries, and the Gas Technology Institute	3 kW PEMFC	Des Plaines, IL				Natural Gas	
Nuvera	RWE Plus AG (Agwill)	5 kW PEMFC of electrical power and 7 kW of heat each	Germany	Apartment buildings throughout	8/02			Part of joint venture between Nuvera and RWE Plus AG to develop and distribute fuel cell systems up to 50 kW (CHP) to Europe Gianfranco.mora@denora.it

Company	Program	Size	Location	Application	Date	Status	Fuel		Notes
Nuvera	MA Technology Collaborative, SatCon Tech. Corp., KeySpan Energy Delivery NE	5 kW telecommunications unit	Massachusetts	Verizon's engineering facility	2001 – 500 hour demonstration	Completed	Natural Gas		US's first fuel cell powered telecommunications site. Mark.a.marchand@verizon.com Derby.r@nuvera.com
Plug Power Inc.	Florida Power and Light Co., DOE	5 kW PEM	North Port, Florida	North Port High School	April 2004	Installed			
Plug Power Inc.	DOD Residential Fuel Cell Demonstration Program Watervliet Arsenal	5 kW (3 units)	Watervliet, NY	Research Facility	Jan. 2002	Installed	Natural Gas		No cogeneration
Plug Power Inc.	DOD Residential Fuel Cell Demonstration Program Watervliet Arsenal	5 kW (3 units)	Watervliet, NY	Manufacturing Facility	Jan. 2002	Installed	Natural Gas		No cogeneration
Plug Power Inc.	DOD Residential Fuel Cell Demonstration Program Watervliet Arsenal	5 kW (4 units)	Watervliet, NY	Officer's Quarters	Jan. 2002	Installed	Natural Gas		No cogeneration
Plug Power Inc.	Plug Power 125 SU-1 Systems delivered in 2001	5 kW each	Various residential locations		2001		Natural Gas		
Plug Power Inc.	Long Island Power Authority (LIPA), DOE, NY State Energy Research and Development Authority (NYSERDA)	6 "alpha" 5 kW CHP units	Four Long Island locations	Homes	2000		Natural Gas		By Feb. 2000, a total of 52 systems were installed in the field to provide power to homes and buildings.

Company	Partner / Authority	System	City	Facility	Date	Status	Fuel	Notes
Plug Power Inc.	Long Island Power Authority (LIPA), DOE, NY State Energy Research and Development Authority (NYSERDA)	12 5 kW units	Various NY	public facilities	2000			An additional 44 systems will be sited over the course of 2001 to complete the 80 unit test, evaluation and demonstration program with: Long Island Power Authority (LIPA, DOE, NY State Energy Research and Development Authority (NYSERDA)
Plug Power Inc.	Long Island Power Authority (LIPA), DOE, NY State Energy Research and Development Authority (NYSERDA)	Three 5 kW units	Brookhaven, NY	DOE's Brookhaven National Lab	March 2000			
Plug Power Inc.	Long Island Power Authority (LIPA), DOE, NY State Energy Research and Development Authority (NYSERDA)	75 PEM systems	West Babylon, NY	LIPA substation	10/2001			The electricity will be distributed to customers through LIPA's electric transmission and distribution system. Expected to produce more than 1 million kWh of electricity during two-year project.
Plug Power Inc.	Long Island Power Authority (LIPA)	5 kW GenSys™ 5C CHP	Babylon, NY	Town Hall	7/2002	On Going		Produces 5 kW electricity and 9 kW of usable.
Plug Power Inc.	Long Island Power Authority (LIPA)	Three 5 kW GenSys™ 5C CHP	Farmingdale, NY	SUNY Farmingdale	2003	Operating	Natural Gas	
Plug Power Inc.	Long Island Power Authority (LIPA)	Three 5 kW GenSys™ 5C CHP	Hempstead, NY	Hofstra University	2002	Operating	Natural Gas	Powering dormitory
Plug Power Inc.	Long Island Power Authority (LIPA)	One 5 kW GenSys™ 5C CHP	Hempstead, NY	Town of Hempstead Animal Shelter	10/24/03	Operating	Natural Gas	
Plug Power Inc.	Long Island Power Authority (LIPA)	Two 5 kW GenSys™ 5C CHP	Garden City, NY	Nassau Community College	2003	Operating	Natural Gas	

Company	Partner	Type	Location	Site	Date	Operating	Fuel	Notes
Plug Power Inc.	Long Island Power Authority (LIPA)	Two 5 kW GenSys™ 5C CHP	Southampton, NY	Southampton College	2003	Operating	Natural Gas	
Plug Power Inc.	Long Island Power Authority (LIPA), DOE, NY State Energy Research and Development Authority (NYSERDA) and National Fuel Gas Distribution #1	5 kW, grid-parallel	Lewiston, New York	Single family home	4/02	Oper. for 1 yr.	Natural Gas	The home is one of two residential fuel cell projects in the service area of National Fuel Gas Distribution Corp. Program will be 1 year long, afterwards the unit will be returned to Plug Power for analysis
Plug Power Inc.	Long Island Power Authority (LIPA), DOE, NY State Energy Research and Development Authority (NYSERDA) and National Fuel Gas Distribution #2	5 kW, grid-parallel	Colden, NY	Single family home	2002	Oper. for 1 yr	Natural Gas	Program will be 1 year long, afterwards the unit will be returned to Plug Power for analysis
Plug Power Inc.	Plug Power's Demonstration Home		Latham, NY		6/98 - 8/15/99 (H2) 12/15/99 - ?? (Natural Gas)		Hydrogen then, Natural Gas	
Plug Power Inc.	DTE Energy Technologies	Two 5 kW PEMFC, grid-parallel	Detroit, MI	Detroit Edison's Hancock Station in Commerce Township, MI	6/2002	Installed; one year project	Natural Gas	Support battery systems that provide backup power to the control circuits for the power station's five natural gas fired turbine peaker units.
Plug Power Inc.	KeySpan Technologies, Inc. and GE MicroGen	30 5 kW	Throughout Long Island and NY metropolitan area				Natural Gas	Agreement made to purchase and test 30 Plug Power residential units throughout Long Island and NY metropolitan area

Company	Partners	Unit	Location	Site	Date	Status	Fuel		Notes
Plug Power Inc.	Flint Energies and GE Fuel Cell Systems	5 kW PEMFC	Warner Robins, GA	Flint Energies' service center facility	7/2002	On Going	Natural Gas		First FCHA demonstrated.
Plug Power Inc.	ISH Fair, German Gas Association, German Technical Surveillance Organization	One Fuel Cell Heating Appliance (FCHA)	Frankfurt, Germany	ISH Fair exhibition area	March 2001	Completed	Natural Gas		
Plug Power Inc.	German Gas Association, German Technical Surveillance Organization	One CE-Certified Fuel Cell Heating Appliance (FCHA)	Gelsenkirchen, Germany	Multiple family home	12/2001	On Going			It supplies seven families with electricity, room heat, and hot water. Provides 80% of the home's electricity and nearly all of the hot water and room heating requirements.
Plug Power Inc.	German Gas Association, German Technical Surveillance Organization	One CE-Certified Fuel Cell Heating Appliance (FCHA)	Essen, Germany	Multiple family home	2001	On Going			Part of a demonstration program funded by the German State of North-Rhine Westphalia and supported by E.ON, Ruhrgas and several local utilities.
Plug Power Inc.	German Gas Association, German Technical Surveillance Organization	One CE-Certified Fuel Cell Heating Appliance (FCHA)	Dusseldorf, Germany	Microbrewery	2001	On Going			Part of a demonstration program funded by the German State of North-Rhine Westphalia and supported by E.ON, Ruhrgas and several local utilities.
Plug Power Inc.	Miller Burton Homes, Built Green CO, City and County of Denver, Governor's OEMC, IREA, Plug Power, Xcel Energy	5 KW CHP residential unit	Denver, CO	Miller Burton Homes' Roaring Fork Parade Home	2002	Parade of Homes: Aug. 31 to Oct. 6 2002	Natural Gas		After the Parade of Homes, this unit will be stationed at the City and County of Denver's fire station at Washington Park, where is will provide a sign. Part of the facility's power and some domestic hot water

Company	Partner/Sponsor	System	Location	Facility	Date	Status	Fuel	Image	Comments
Plug Power Inc.	Governor's Office of Energy Management and Conservation (OEMC), Plug power, Xcel Energy, & City and County of Denver	5 kW CHP residential unit	Denver, CO	Washington Park Fire Station	12/2002	Dec. 2002 – Dec. 2003	Natural Gas		This is the same unit displayed in the Parade of Homes. This demonstration will last one year. This unit provides a portion of the facilities' electricity and heat.
Plug Power Inc.	Plug Power Long Island Power Authority (LIPA)	Three 5 kW Back up UPS system	Kings Point, New York	US Merchant Marine Academy	To be installed Fall 2002/3		Hydrogen		The new system marks Plug Power's first shipment into the backup/UPS markets. LIPA will install the units.
Plug Power Inc.		9 5 kW GenSys™ 5C CHP PEMFCs	Naval facilities in California	See Comments	9/2002	On Going	Natural Gas		8 will power & heat bachelor-enlisted quarters, a gym, and laundry facility at the Naval Air Station North Island and the Submarine Base located in San Diego, Ca. The ninth system will power & heat an indoor pool at the Naval Air Weapons Station in China Lake, Ca. Represents the first Plug Power systems installed and operated in California
Plug Power Inc.	Bonneville Power Administration Northwest Natural Inc.	5 kW GenSys CHP System	Hillsboro, OR	Harkins House Juvenile Detention Facility	Nov 2003	Installed			Mira Vowles 503.230.4796 mkvowles@bpa.gov
ReliOn (formerly Avista Labs)	Avista Labs and US Army Corps of Engineers Buy Down Program	3 kW PEMFC (SR-72)	Washington Airport at Geiger Field in Spokane, WA	Guard facility	2002	On Going	Industrial Grade bottled Hydrogen		

Company	Program	Size	Location	Facility	Date	Commissioned/Installed	Fuel	Image	Notes
ReliOn	DOD Residential Fuel Cell Demonstration Program MCB Kaneohe Bay	5 kW (1 unit)	Kaneohe Bay, HI	TBA	Dec. 2002	Not installed	Propane		No cogeneration
ReliOn	DOD Residential Fuel Cell Demonstration Program Ft. Jackson	5 kW (1 unit)	Columbia, NC	Base Housing	Dec. 2002	Not installed	Natural gas		No cogeneration
ReliOn	DOD Residential Fuel Cell Demonstration Program Barksdale AFB	5 kW (1 Unit)	Bossier City, LA	Base Housing	Dec. 2002	Not installed	Natural gas		No cogeneration
ReliOn	FCTEC http://www.fctec.com/main.html (DOD)	720 W PEM Avista Fuel Cell	Johnstown, PA	FCTEC					Test Program Mr. Scott Kenner at *CTC.* Phone (814) 269-2891 kenners@ctc.com
ReliOn	Bonneville Power Administration	1 kW		Central Washington University	January 2004	Commissioned			
Siemens Westinghouse	Siemens and Edison SpA	300 kW pressurized system (PH) SOFC CHP	Milan, Italy (Sinetta Marengo)	Edison SpA – research facility	Scheduled for 9/02		PNG		Cell length = 1500 mm / 1728 cells/stack
Siemens Westinghouse	EPA at Fort Meade	1 MW pressurized (PH) Hybrid SOFC system	Fort Meade: Just outside Washington, DC	Fort Meade EPA Lab	2003		Natural Gas		Cinergy, with headquarters in Cincinnati, Ohio, will be the operator of this demonstration. Cell length = 1500 mm / 5760 cells/stack

Manufacturer	Project	System	Location	Consortium	Date	Status	Fuel		Notes
Siemens Westinghouse	RWE	300 kW SOFC/gas turbine pressurized (PH) hybrid CHP system	Essen, Germany	RWE Energie Meteorit Exposition	Scheduled for 4/02	In Development	Natural Gas		Zero SOx, and less than 1 ppm NOx emissions expected. Cell length = 1500 mm / 1728 cells/stack. Expected electrical efficiency = 58%
Siemens Westinghouse	Stadtwerke Hanover AG and E.ON Energie AG	250 kW SOFC	Hanover, Germany	Herrenhausen power plant	2003				
Siemens Westinghouse	CO_2 capture demonstration with Shell	250 kW SOFC hybrid system	Norway		2003	In Development	Natural Gas		This unit is being fabricated for Shell in Norway to demonstrate the principle of CO_2 capture from power plants. By keeping the fuel and air streams separate, the CO_2 generated can be captured, and the SOFC power system will essentially have clean water as its only effluent. Cell length = 1500 mm / 2304 cells/stack
Siemens Westinghouse	Pre-commercialization with EDB/ELSAM	Siemens Westinghouse 125 kW SOFC/GT hybrid co-generation system	Westervoort, Netherlands (near Arnhem)	EDB/ELSAM – a consortium of Dutch and Danish utilities / power to the local grid	Dec. 1997		Natural Gas		It is believed to be the highest sustained efficiency of any simple fuel cell system operating on natural gas. It also contributed 64 kW of heat into the district's hot water system. Put out average of 63 kW thermal output as hot water into the districts hot water system. Output of negligible NOx and SOx emissions. 1 stack; 1500 mm long; 1152 cells per stack

Manufacturer	Partner/Customer	System	Location	Site owner	Date	Status	Fuel	Image	Notes
Siemens Westinghouse	At EDB/ELSAM, a consortium of Dutch and Danish utilities	Siemens Westinghouse SOFC/GT hybrid co-generation system	Westervoort, Netherlands		Dec. 1999	Unit Moved to RWE (below)	PNG		This system typically supplies 109 kW electrical power and 63 kW thermal output as hot water. As of November 2000, this system had operated for over 16,000 hours achieving 46% electrical eff. at 109 kW net AC power. Operated over 12,577 hours
Siemens Westinghouse	RWE	Siemens Westinghouse 100 kW SOFC co-generation system	Germany	RWE	2001	Still in Testing (moved from EDB/ELSAM)	PNG		1 stack; 1500 mm long; 1152 cells per stack. Operated over 4000+ hours
Siemens Westinghouse	EC and DOE	1 MW pressurized (PH) hybrid SOFC system,	Marbach, Germany	Energie Baden-Wurttemberg (EnBW)'s site	November, 2003		PNG		A consortium of European utilities led by ENERGIE will be the Program Manager. This will be the largest fuel cell CHP demonstration in Europe. Cell length = 1500 mm / 5760 cells/stack
Siemens Westinghouse	JGU-1	20 kW SOFC/GT hybrid system			1992	Finished	PNG (Pipe line Natural Gas)		2 stacks, 500 mm long, 576 cell per stack. Logged 817 hours
Siemens Westinghouse	JGU-2	25 kW SOFC/GT hybrid			1995	Finished	PNG		1 stack; 500 mm long; 576 cells per stack. Logged 13294 hours
Siemens Westinghouse	OPT	250 kW SOFC/GT hybrid system			2002	Planned	PNG		Cell length = 1500 mm / 2304 cells/stack

Siemens Westinghouse	Osaka Gas	3 kW SOFC/GT hybrid		1987	Finished	H_2+CO		1 stack 360 mm long with 144 cells; Logged 3012 hours
Siemens Westinghouse	South California Edison PSOFC/MTG	180 kW PSOFC/GT hybrid system		2000	Finished	Natural Gas		1 stack: 1500 mm long; 1152 cells per stack; Logged 770 hours
Siemens Westinghouse	South California Edison	20 kWe SOFC/GT hybrid system		1994	Finished	Natural Gas		1 stack: 500 mm long; 576 cells per stack; Logged 6015 hours
Siemens Westinghouse	South California Edison	27 kW SOFC/GT hybrid system		1995	Finished	Natural Gas; Diesel; Jet fuel		1 stack: 500 mm long; 576 cells per stack; Logged 5582 hours
Siemens Westinghouse	Tokyo Gas	3 kW SOFC/GT hybrid system		1987	Finished	H_2+CO		1 stack 360 mm long with 144 cells; Logged 4882 hours
Siemens Westinghouse	TVA	0.4 kW SOFC/Gas Turbine hybrid power systems		1986	Finished	H_2+CO		1 stack 300 mm long with 24 cells; Logged 1760 hours
Siemens Westinghouse	Utilities-A	20 kW SOFC/GT hybrid system		1992	Finished	PNG		1 stacks; 500 mm long; 576 cells per stack; Logged 2601 hours
Siemens Westinghouse	Utilities-B1	20 kW SOFC/GT hybrid system		1992	Finished	PNG		1 stack: 500 mm long; 576 cells per stack; Logged 1579 hours
Siemens Westinghouse	Utilities-B2	20 kW SOFC/GT hybrid system		1993	Finished	PNG		1 stack: 500 mm long; 576 cells per stack; Logged 7064 hours

Company	Program/Description	System	Location	Site	Date	Status	Fuel		Notes
Siemens Westinghouse	NFCRC under contract w/ Edison International and Siemens Westinghouse Power Corp. w/ funding from DOE and Ca Energy Commission	220 kW SOFC w/ micro-turbine generator hybrid	University of California (Irvine, CA)	NFCRC		Has started its test period, scheduled for 3000 hours	Natural Gas		World's first Fuel Cell-Gas micro-Turbine Hybrid. The micro-turbine is driven by the hot pressurized exhaust produced by the fuel cells. Set a new efficiency record for a fuel cell system operating on natural gas. Paul Klein, Southern California Edison: 626-302-8066 Drew Malcomb, DOE office of Public Affairs, 202-586-5806
Siemens Westinghouse	Prototype Demonstrations with BP America, Inc. Buy Down Program Recipient FY2000	250 kW SOFC CHP co-generation system	Nikiski, Alaska	Gas-to-liquids plant	2003		Natural Gas		
Siemens Westinghouse	Prototype Demonstrations with Kinectrics Buy Down Program Recipient FY1999	250 kW SOFC atomospheric pressure system (ATM)	Toronto, Canada	Kenectrics (formerly Ontario Hydro)	Mid-2002				Next step in Siemens's Commercialization plan is to demonstrate prototypes of the first commercial products in 2001-2004
Stuart Energy Systems	South Coast Air Quality Management District	120kW	Diamond Bar, CA	AQMD headquarters	Nov 2003	To be delivered early 2004	Hydrogen		Stuart Energy Media and Public Relations Wanda Cutler 905-282-7769 Stuart Energy Investor Relations 905-282-7727

Sulzer Hexis	E.ON Energie AG, München	1 kW HXS 1000 PREMIERE			Natural Gas	www.eon-energie.de One of Europe's largest private electricity, gas and water utilities purchase a total of 55 «HXS 1000 PREMIERE» over the period 2002-03.
Sulzer Hexis	EnBW Energie Baden-Württemberg AG, Karlsruhe	1 kW HXS 1000 PREMIERE			Natural Gas	www.enbw.de One of the largest German supraregional utilities will purchase a total of 40 systems «HXS 1000 PREMIERE» over the period 2001 – 03.
Sulzer Hexis	EWE PLC, Oldenburg	1 kW HXS 1000 PREMIERE			Natural Gas	www.ewe.de The largest regional utility in Germany will purchase a total of 155 Hexis fuel cells systems «HXS 1000 PREMIERE» over the period 2001 – 03.
Sulzer Hexis	EWR Elektrizitätswerk Rheinhessen AG, Worms	1 kW HXS 1000 PREMIERE			Natural Gas	www.ewr.de The utility for the region Rheinhessen/Ried will purchase a total of 60 «HXS 1000 PREMIERE» over the period 2001 – 03.
Sulzer Hexis	Thyssengas GmbH, Duisburg	1 kW HXS 1000 PREMIERE			Natural Gas	www.thyssengas.de One of the important natural gas importer and supplier will purchase a total of 42 «HXS 1000 PREMIERE» over the period 2002-03.
Sulzer Hexis	VNG - Verbundgasnetz AG, Leipzig	1 kW HXS 1000 PREMIERE			Natural Gas	www.vng.de The East German utility for gas piped over a long distance purchase a total of 16 «HXS 1000 PREMIERE» over the period 2002-03.

					Manufact. Of all systems completed by end of 2003	Manufacturing		
Sulzer Hexis	GVM	30 residential systems	Throughout Germany, Austria, and Switzerland	Various GVM customer homes			Natural Gas	This part of Sulzer's agreement to place 400 pre-series FC systems. Sulzer delivered 58 systems in 2002
Tennessee Valley Authority					Nov 2001	Begin operation in 2003		
Teledyne (formerly Energy Partners)	DOE	3 kW fuel cell distributed generator	Columbus, Mississippi		2001	Completed Fall 2001	Natural Gas	Field testing of residential natural gas systems started in 2001
Teledyne (formerly Energy Partners)	DOE	7 kW distributed generator			2002	To be released to DOE in Fall 2002		Successor unit of the 3 kW
UTC Fuel Cells	EPA in Ann Arbor, MI. Noresco	PC25 200 kW PAFC	Ann Arbor, MI	EPA building	January 2004	Operational	Natural Gas	Grid Parallel setup. Manufactured in 1996, this unit had numerous upgrades installed prior to its installation here. It has been limited to 175Kw output due to overheating internally. UTC believes the new stack and heat exchangers will allow the output to increase to 200Kw. Apparently, a leaking heat exchanger caused the premature failure of one of the stacks. Noresco, LLC, contact Rick Levin 508-614-1006 Steven V. Dorer, Facilities Manager: dorer.steven@epa.gov 734-214-4503

Company	Program	Model	Location	Application	Date	Status	Fuel		Contact / Notes
UTC Fuel Cells	FCTEC http://www.fctec.com/main.html (DOD)	PC25C 200 kW PAFC	Johnstown, PA	FCTEC	January 1999	On going until 1/2003	Natural Gas		Bill Taylor w-taylor@cecer.army.mil Mr. Bob Unger at CTC. Phone: (814) 269-2721 ungerb@ctc.com
UTC Fuel Cells	PAFC Demonstration Program (DOD) Fort Bliss	Model PC25C Grid connected at new transformer	El Paso, TX	Laundry	Sep.-1997	Operational	Natural Gas		Thermal output heats laundry hot water storage tanks Total estimated thermal utilization ~17% Mr. Joe Mathis (915) 568-3107/6627 Commander Directorate of Public Works Attn: ATZC-ISE-P (Joe Mathis) Ft. Bliss, TX 79916-0058
UTC Fuel Cells	PAFC Demonstration Program (DOD) Fort Eustis	Model PC25B	Newport News, VA	Gymnasium/Pool	1/2004	Project Completed Successfully	Natural Gas		Thermal output to DHW and pool (~68% est. thermal utilization) Mr. Daniel Wood (757) 878-2489 ext. 228 woodd@eustis.army.mil Daniel.benito.wood@us.army.mil Directorate of Public Works Building 1407 Attn: ATZF-PWW (Daniel Wood) Ft. Eustis, VA 23604

UTC Fuel Cells	UTC and TBE GmbH with GEW Koln AG	UTC PAFC 220 kW	Cologne, Germany	Waste treatment plant			Digester Gas		The first time in Europe a fuel cell was used to utilize waste methanol produced from sewage to generate electricity and heat efficiency. Heat generated will be used in the sewage treatment process. This supplements 4 other systems operated by UTC/ONSI at waste water treatment plants in USA, 3 in Japan, and systems at nearly 200 other locations worldwide
UTC Fuel Cells	First National Bank of Omaha and Sure Power Corp.	Four PC25 200 kW systems	Omaha, NE	National Bank	Nov. 1999	On Going (purchase)	Natural Gas		
UTC Fuel Cells	Hospital in Staten Island, NY	200 kW PAFC	Staten Island, NY	Hospital			Natural Gas		
UTC Fuel Cells	Hotel in Spokane, WA	200 kW PAFC	Spokane, WA	Hotel			Natural Gas		
UTC Fuel Cells	Las Virgenes Municipal Water District (AGD) in Calabasas, CA	200 kW PAFC	Calabasas, CA						

Company	Program	Model	Location	Site	Date	Status	Fuel	Image	Notes / Contact
UTC Fuel Cells	PAFC Demonstration Program (DOD) Fort Huachuca	Model PC25C Grid connected at existing electrical transformer (no emergency back-up)	Sierra Vista, AZ	Riley Barracks	Jul.-1997	Operational	Natural Gas		High grade thermal output (option) to space heating loop; Thermal output to domestic hot water (total ~44% est. thermal utilization). Mr. Bill Stein (520) 533-1861; U.S. Army Intelligence Center & Fort Huachuca; Attn: ATZS-EHE (Bill Stein); Building 22422; Fort Huachuca, AZ 85613-6000
UTC Fuel Cells	PAFC Demonstration Program (DOD) Fort Richardson	Model PC25C Grid connected at existing electrical panel (no emergency back-up)	Anchorage, AK	National Guard Armory	Apr.-1997	Off-line, Apr-2001	Natural Gas		High grade thermal output (option) to space heating loop; Thermal output to domestic hot water (total ~45% est. thermal utilization). Tom C. LaVictoire 907-428-6780; Dept. of Military and Veteran Affairs; P.O. Box 5-549; Camp Denali; Fort Richardson, AK 99505
UTC Fuel Cells	PAFC Demonstration Program (DOD) U.S. Army Soldier Systems Center	Model PC25B Grid connected at existing sub panel	Natick, MA	Boiler Plant	Feb.-1995	Operational	Natural Gas		Thermal output to storage tank (~45% est. thermal utilization). Mr. David Duncan, P.E. (508) 233-4934; Commander; U.S. Army; Soldier Systems Center; Kansas Street; Attn: AMSSC-S-PWTD (David Duncan, P.E.); Natick, MA

Manufacturer	Program / Site	Description	Location	Application	Startup	Status	Fuel	Image	Notes / Contact
UTC Fuel Cells	PAFC Demonstration Program (DOD) Picatinny Arsenal	Model PC25B Grid connected at panel inside electric room	Dover, NJ	Boiler Plant	Oct.-1995	Shut down, Jul-2001	Natural Gas		Thermal output preheats make-up water (~100% est. thermal utilization) Mr. Richard Sloboda (201) 724-2492 Commander ARDEC, Picatinny Arsenal Building 506 ASTMA-AR-PW (Richard Sloboda) Picatinny Arsenal, NJ 07806-5000
UTC Fuel Cells	PAFC Demonstration Program (DOD) Pine Bluff Arsenal	Model PC25B Grid connected at splice after pole mounted transformer Grid independent terminals power the boiler plant.	White Hall, AR	Boiler Plant	Oct.-1997	Shut down, Jan-2000	Natural Gas		Thermal output heats boiler make-up water Total estimated thermal utilization ~90% Ms. Nancy Rimmer (870) 540-3312 Commander Pine Bluff Arsenal Attn: Nancy Rimmer Building 34980 Pine Bluff Arsenal, AR 71602-9500
UTC Fuel Cells	PAFC Demonstration Program (DOD) US Military Academy	Model PC25B Grid connected at existing panel	West Point, NY	Central Boiler Plant	Dec.-1995	Decommissioned, Feb-2001	Natural Gas		Thermal output for boiler make-up water (~70% est. thermal utilization) Mr. Bob Kronk (914) 938-6873 Commander U.S. Military Academy DHPW MAEN-U-N (Bob Kronk) Building 667 West Point, NY 10996

Company	Program	Model/Details	Location	Facility	Date	Status	Fuel	Image	Notes/Contact
UTC Fuel Cells	PAFC Demonstration Program (DOD) Watervliet Arsenal	Model PC25B Grid connected at existing electrical panel Emergency back-up for grid-independent operation	Albany, NY	Central Boiler Plant	Oct.-1997	Operational	Natural Gas		Thermal output to preheat boiler make-up (~58% est. thermal utilization) Mr. Paul Gentiluomo (518) 266-3680 gentiluomo@wva.army.mil Watervliet Arsenal Attn: SIOWV-PW (Paul Gentiluomo) Building 120 Watervliet Arsenal, NY 12189-4050
UTC Fuel Cells	PAFC Demonstration Program (DOD) 911th Airlift Wing Consolidated Natural Gas	Model PC25C Grid connected (no emergency back-up)	Pittsburgh, PA	Central Heating Plant	Feb.-1997	Off-line, Sep-2001	Natural Gas		Mr. Jay Hague (412) 474-8573 911th Airlift Wing/CE Pittsburgh IAP ARS Attn: Jay Hague 1113 Herman Avenue Coraopolis, PA 15108
UTC Fuel Cells	US Airways, Peoples' Natural Gas	Model PC25C 200 kW	Pittsburgh, PA	U.S. Air's hanger #2 @ Pittsburgh International Airport			Natural Gas		
UTC Fuel Cells	PAFC Demonstration Program (DOD) 934th Airlift Wing	Model PC25B Grid connected at new electrical transformer (fuel cell option)	Minneapolis, MN	Boiler Plant	Feb.-1995	Off-line, Sep-2000	Natural Gas		Thermal output to preheat boiler make-up water (~45% est. thermal utilization) Mr. Mehrdad "Dodd" Sadeghi (612) 713-1912 934th Airlift Wing Attn: Mehrdad "Dodd" Sadeghi SPTG/CEC 760 Military Highway Minneapolis, MN 55450-2000

UTC Fuel Cells	PAFC Demonstration Program (DOD) Barksdale AFB	Model PC25C Grid connected at panel in electrical room	Bossier City, LA	Hospital	Jul-1997	Operational	Natural Gas		Thermal output heats hydronic space heat/reheat loop. Total estimated thermal utilization ~90% Mr. Nathaniel B. Cost (318) 456-5039 Nathan.Cost@BARKSDALE.AF.MIL Nathaniel B. Cost, E.I., C.E.M. Electrical Engineer/Energy Manager 2 CES/CEOE 334 Davis Ave. W. #200 Barksdale AFB, LA 71110
UTC Fuel Cells	PAFC Demonstration Program (DOD) Davis-Monthan AFB	Model PC25C Grid connected at new transformer (Program option)	Tucson, AZ	Gymnasium	Dec.-1997	Operational	Natural Gas		High grade thermal output (Program option) to absorption chillers. Low grade thermal output to hot water storage tank. Total estimated thermal utilization ~65% Mr. John Li (520) 228-3203 Davis-Monthan AFB Attn: John Li 355 CES/CEOE 3791 S. Third Street Tucson, AZ 85707

Manufacturer	Program	Model / Grid	Location	Application	Date	Status	Fuel	Image	Notes / Contact
UTC Fuel Cells	PAFC Demonstration Program (DOD) Edwards AFB	Model PC25C Grid connected at existing transformer	Palmdale, CA	Hospital	Jul.-1997	Operational	Natural Gas		High grade thermal output (Program option) to space heating loop. Total estimated thermal utilization ~23%. **Lt. Sufnar (661) 277-7021, x1537** Attn: Lt. Sufnar, Edwards AFB, 225 North Rosamond, Edwards AFB, CA 93534
UTC Fuel Cells	PAFC Demonstration Program (DOD) Kirtland AFB	Model PC25B Grid connected at switch tied to electrical transformer. Grid independent connection to entire boiler plant	Albuquerque, NM	Boiler Plant	Sep.-1995	Decommissioned, Dec.-2001	Natural Gas		Thermal output to deaerator tank (~55% est. thermal utilization). Mr. Cliff Richardson (505) 846-7979. Mr. Cliff Richardson, 2050 Wyoming Blvd., 877 CES/CEOE, Kirtland Air Force Base, NM 87117-5663
UTC Fuel Cells	PAFC Demonstration Program (DOD) Laughlin AFB	Model PC25C Grid connected at existing electrical transformer (no emergency back-up)	Del Rio, TX	Hospital	Sep.-1997	Operational	Natural Gas		Thermal output to space heat/cool reheat loop and DHW loop (~75% est. thermal utilization). Lt. Dana Repak (830) 298-5960. Laughlin AFB, Attn: Lt. Dana Repak, 47 CES/CEOE, 251 Fourth Street, Laughlin AFB, TX 78843

Program	Model	Location	Facility	Date	Status	Fuel	Image	Notes
PAFC Demonstration Program (DOD) Little Rock AFB	Model PC25C Grid connected at electrical panel	Jacksonville, AR	Hospital	Oct.-1997	Shut down, Dec.-2000	Natural Gas		Thermal output heats space conditioning recirculation loop. Total estimated thermal utilization ~85%. Darin Bailey (501) 987-6704 Little Rock AFB Attn: Darin Bailey 314th CES/CEOE 536 Thomas Avenue Little Rock, AR 72099-5005
PAFC Demonstration Program (DOD) Nellis AFB	Model PC25B Grid connected at main breaker panel	Las Vegas, NV	Central Plant for Dorm Facility	Oct.-1995	Off-line, Jun-2001	Natural Gas		Currently waiting transfer to another DOD facility. Thermal to make-up water and heat pump loop (~40% est. thermal utilization). Mr. Gary Kaiser (702) 652-7790 Nellis AFB Attn: Gary Kaiser 99CES/CE2 6020 Beale Ave Nellis AFB, NV 89191
PAFC Demonstration Program (DOD) Vandenberg AFB	Model PC25A	Lompoc, CA	Space Control Center	March 25, 1994	Decommissioned, Feb-2001	Natural Gas		The power plant was moved to Pinal Air Park, Arizona Air National Guard in Tucson, Arizona and began operation on March 26, 2001. It is now operating at 175 kW to facilitate long-term operation of the power plant.

Program	Model/Configuration	Location	Application	Date	Status	Fuel	Image	Notes / Contact
PAFC Demonstration Program (DOD) Westover ARB	Model PC25C Grid connected at new electrical transformer (Program option)	Chicopee, MA	Boiler Plant	Sep.-1997	Operational	Natural Gas		Low grade thermal output heats boiler make-up water. High grade thermal output (Program option) to condensate return loop. Total estimated thermal utilization ~45% est. Jon Wyman Wk: (413) 557-3575 Westover Air Reserve Base 439 SPTG/CECD 250 Patriot Avenue, Suite 1 Chicopee MA 01022-1670
PAFC Demonstration Program (DOD) CBC Port Hueneme	Model PC25B Grid connected at new transformer	Port Hueneme, CA	Swimming Pool	Aug.-1997	Off-line, Dec-2001	Natural Gas		Thermal output heats swimming pool. Total estimated thermal utilization ~92%. Mr. Carl Rhoads (805) 982-4313 rhoadsci@bcph.navy.mil CBC Port Hueneme Attn: Carl Rhoads 1000 23rd Avenue Code PW210, Bldg 850 Port Hueneme, CA 93043
PAFC Demonstration Program (DOD) Naval Hospital MCB Camp Pendleton	Model PC25B Grid connected at existing panel	Oceanside, CA	Naval Hospital	Oct.-1995	Decommissioned, Jan.-2001	Natural Gas		Thermal output for DHW storage (~75% est. thermal utilization) Mr. Jim Beesing (760) 725-5447 Naval Hospital Attn: Jim Beesing Facilities Management Department Code 01E.Jb, Box 555191 Camp Pendleton, CA 92055

Manufacturer	Program	Model/Connection	Location	Application	Date	Status	Fuel	Image	Notes/Contact
UTC Fuel Cells	PAFC Demonstration Program (DOD) NAS Fallon	Model PC25C Grid connected at new electric transformer (fuel cell option) Grid independent connection at new electric transformer (fuel cell option)	Fallon, NV	Galley	Mar.-1997	Operational	Natural Gas		Thermal output to DHW loop (~10% est. thermal utilization) David Gisi (702) 426-2410 NAS Fallon Public Works Engineering Building 307 Fallon, NV 89496
UTC Fuel Cells	PAFC Demonstration Program (DOD) Naval Hospital NAS Jacksonville Logan Energy	Model PC25C Grid connected at existing electrical panel (no emergency back-up)	Jacksonville, FL	Naval Hospital	Apr.-1997	On Going	Natural Gas		Thermal output to DHW loop (~56% est. thermal utilization) Mr. Larry E. Forbes (904) 777-7593 Larry E. Forbes Facilities Management Dept (Code 0104) Naval Hospital 2080 Child Street Jacksonville, FL 32214-5104
UTC Fuel Cells	PAFC Demonstration Program (DOD) Naval Station Newport	Model PC25PC25 B Grid connected at boiler plant electrical transformer	Newport, RI	Boiler Plant	Feb.-1995	Shut down, Apr-2001	Natural Gas		Thermal output to preheat boiler make-up water (~60% est. thermal utilization) Mr. John Alfano (401) 841-2161 Newport NETC Attn: John Alfano Engineering Div. Code 422 Public Works Dept., One Simonpietri Drive Newport, RI 02841-1711

UTC Fuel Cells	PAFC Demonstration Program (DOD) Naval Oceanographic Office	Model PC25B Grid connected at electrical panel Grid independent load connected at new panel	Stennis Space Center, MS	Naval Oceanic Center for NAVO support	Sep.-1997	Operational	Natural Gas		Thermal output used for space heat/reheat loop Total estimated thermal utilization ~12% Mr. Mike Killam (228) 688-4062 killamm@navo.navy.mil US Naval Oceanographic Office Attn: Mike Killam 1002 Balch Boulevard Stennis Space Center, MS 39522-5001
UTC Fuel Cells	PAFC Demonstration Program (DOD) Subase New London	Model PC25C Grid connected at existing electrical panel	Groton, CT	Boiler Plant	Oct.-1997	Operational	Natural Gas		Thermal output heats boiler make-up water Total estimated thermal utilization ~90% John Papitto Wk: (860) 694-4914 Naval Submarine Base New London Attn: Code 8243 Engineering Division Box 400, Building 135 Groton CT 06349

Company	Program	Model / Connection	Location	Application	Date	Status	Fuel	Image	Notes / Contact
UTC Fuel Cells	PAFC Demonstration Program (DOD) Naval Hospital MCAGCC Twenty-nine Palms	Model PC25B Grid connected at existing sub panel Grid independent connection at new electrical sub panel	Twenty-nine Palms, CA	Naval Hospital	Jun.-1996	Decommissioned, May-2000	Natural Gas		Thermal output to DHW loops (~60% est. thermal utilization) Mr. Patrick Dougherty (760) 830-2395 tnp1pmd@tnp20.med.navy.mil Naval Hospital, 29 Palms MCB Attn: Facilities Manager (Patrick Dougherty) MCAGCC, Box 788250 29 Palms MCB, CA 92278-8250
UTC Fuel Cells	PAFC Demonstration Program (DOD) US Naval Academy	Model PC25C Grid connected in electrical room	Annapolis, MD	Galley	Sep.-1997	Operational	Natural Gas		Thermal output heats make-up water Total estimated thermal utilization ~78% Mr. Chi Chiu (410) 293-1091 chiu@arctic.nadn.navy.mil US Naval Academy Attn: Chi Chiu Public Works Department (Code 150) 181 Wainwright Road Annapolis, MD 21402-5012
UTC Fuel Cells	PAFC Demonstration Program (DOD) National Defense Center for Environmental Excellence (NDCEE)	Model PC25C Grid connected in spare panel slot	Johnstown, PA	Industrial Building	Aug.-1997	Operational	Natural Gas		High grade thermal output heats evaporator tank Total estimated thermal utilization ~19% Bob Lentz (814) 269-6456 lentz@ctc.com National Defense Center For Environmental Excellence Attn: Bob Lentz Concurrent Technologies Corporation 1450 Scalp Avenue Johnstown.

Company	Project	Type	Location	Application	Year	Status	Fuel	Image	Description
UTC Fuel Cells	Australian Technology Park Sydney LTD Buy Down Program Recipient FY1998	200 kW PAFC	Sydney, Australia	Technology Park	FY1998		Natural Gas		Australia's First Fuel Cell.
UTC Fuel Cells	Toshiba, Institute of Energy Economics of Japan, New Energy and Industrial Technology Development Organizations (NEDO) of Japan	200 kW PC25™	Guangzhou (Canton) City, China	Hog farm	Late 2001	On going	Liquefied petroleum gas (LPG) and then waste methane gas produced at farm		First commercial fuel cell power installation in China. Sold to Toshiba Corp. which was modified and sold to the customer. The unit is managed by the Industrial Technology Development Organization (NEDO) of Japan. At this point (1/9/02), UTC has delivered more than 225 fuel cells to customers in 19 countries on five continents.
UTC Fuel Cells	AEC South County Hospital	200 kW PAFC		Hospital	FY1998		Natural Gas		
UTC Fuel Cells	South County Hospital	200 kW PAFC	Wakefield, RI	South County Hospital			Natural Gas		
UTC Fuel Cells	Alcorn State University Logan Energy	200 kW PAFC	Alcorn, Mississippi	Supporting Alcorn State University's campus grid	9/2000	On Going	Natural Gas		

Manufacturer	Customer	System	Location	Facility	Date	Status	Fuel	Image	Notes
UTC Fuel Cells	Assured Power	UTC PC25 200 kW system					Natural Gas		
UTC Fuel Cells	Braintree Electric light Department	200 kW PAFC	Braintree, MA		FY1995				Charles R. Berry (718) 403-3065 — 1995 Cogeneration Project of the Year by the Cogeneration and Competitive Power Institute
UTC Fuel Cells	Brooklyn Union Gas Co.	200 kW PAFC	Brooklyn, NY	Saint Vincent's Hospital Laundry facility	(Pre-Commercial) Oct. 1992	On Going	Natural Gas		
UTC Fuel Cells	Chevron	200 kW PAFC			FY2000				
UTC Fuel Cells	Chugach Electric Association — Buy Down Recipient FY1998	Five 200 kW PAFC systems	Anchorage, Alaska	US Postal Service headquarters	2000	On Going	Natural Gas		In 2000, it was the largest commercial fuel cell system in the nation, comprising of 5 UTC PC25 systems. It is also the first time a fuel cell system is part of an electric utility's grid
UTC Fuel Cells	City of Mesa Utilities Department — Buy Down Program Recipient FY1995	200 kW PAFC							
UTC Fuel Cells	City of Portland, OR — Buy Down Recipient FY1996-1997	200 kW PAFC system	Portland, OR	Columbia Blvd. Waste Water Treatment Plant (ADG)	May 1999		Digester Gas		

UTC Fuel Cells	CLC S.r.l. Italy Ansaldo/Energie und Wasserversorgung AG Buy Down Recipient FY1995	200 kW PAFC		Installed 1996/7	Natural Gas		Ansaldo (ARI) is also the exclusive dealer of International Fuel Cells (IFC) for Italy, France and Spain (non-exclusive for the rest of Europe) for the PC25C fuel cells powerplant. Ansaldo's next unit was converted to run on liquid hydrogen and was placed next to the original Natural Gas fed one as a 2 year demonstration. http://www.ansaldoricerche.it /clc/ari.htm
UTC Fuel Cells	CLC S.r.l. Italy Ansaldo / Sun Chemical Buy Down Recipient FY1995	Four 200 kW PAFC systems	Sun Chemical		Natural Gas		
UTC Fuel Cells	CLC S.r.l. Italy Ansaldo/Vattenfall AB Buy Down Recipient FY1995	200 kW PAFC	Vattenfall AB		Natural Gas		
UTC Fuel Cells	CLC Srl Italy Ansaldo/Stadtwerke Oranienburg Buy Down Recipient FY1995	200 kW PAFC	Stadtwerke Oranienburg		Natural Gas		

Company	Project	System	Location	Customer	Year	Status	Fuel	Image	Description
UTC Fuel Cells	CLC Srl Piazza Carignano 2, Bu Down Recipient FY1995	200 kW PAFC					Natural Gas		
UTC Fuel Cells	C.l.C.r.l. Italy Ansaldo/Gasversorgung Sachen Anhalt Gmbh, Buy Down Recipient FY1995	200 kW PAFC					Natural Gas		
UTC Fuel Cells	C.l.C.r.l. Italy Ansaldo/Stadtwerke Saarbrucken AG, Buy Down Recipient FY1995	200 kW PAFC		Stadtwerke Saarbrucken AG			Natural Gas		
UTC Fuel Cells	Connecticut Juvenile School, Buy Down Recipient FY1996-1997	Six 200 kW PAFC systems (1.2 MW)	Middletown, Connecticut	Juvenile Training School	2001	On Going	Natural Gas		These are used in conjunction with traditional generators and the national grid to provide power, heating and cooling to the facility.
UTC Fuel Cells	Connecticut Natural Gas Corp., Bu Down Recipient FY1995	200 kW PAFC	Connecticut	Natural Gas Corp.			Natural Gas		
UTC Fuel Cells	CTG Corporation, Buy Down Recipient FY1996-1997	Two 200 kW PAFC systems	Uncasville, Connecticut	Mohegan Sun Casino Hotel		On Going	Natural Gas		

Company	Recipient	System	Location	Building	Status	Fuel	Image	Notes
UTC Fuel Cells	Durst Corporation Buy Down Recipient FY1996-1997	Two 200 kW PAFC systems	4 Times Square, New York	4th floor Conde Nast Building	On going	Natural Gas		Provides power for the NASDAQ sign
UTC Fuel Cells	Energy 2000 Buy Down Recipient FY1996-1997	Two 200 kW PAFC systems		Las Virgines Wastewater Treatment		Digester Gas		
UTC Fuel Cells	Equitable Resources Buy Down Recipient FY1995	200 kW PAFC						Equitable Resources installed 10 PC25s at US Facilities of DOD in 1997
UTC Fuel Cells	Equitable Resources Bu Down Recipient FY1995	200 kW PAFC						
UTC Fuel Cells	Equitable Resources Buy Down Recipient FY1995	200 kW PAFC						
UTC Fuel Cells	Equitable Resources Buy Down Recipient FY1995	200 kW PAFC						

Company	Recipient	Type	Location	Facility	Date	Status	Fuel		Notes
UTC Fuel Cells	Equitable Resources Buy Down Recipient FY1995	200 kW PAFC	Oakmont, PA	Presbyterian Medical Center			Natural Gas		
UTC Fuel Cells	Equitable Resources, NORESCO Buy Down Recipient FY1995	200 kW PAFC							
UTC Fuel Cells	Gas, Elektrizitats und Wasserwerk Buy Down Recipient FY1998	200 kW PAFC							
UTC Fuel Cells	Hamilton Sundstrand Buy Down Recipient FY1995	200 kW PAFC	Windsor Locks, CT	Hamilton Sundstrand Data Center	Installed		Natural Gas		
UTC Fuel Cells	HGC Hamburg Gas Consult Buy Down Recipient FY1998	200 kW PAFC	Hamburg, Germany			On Going			
UTC Fuel Cells	Logan Energy Rebekah Baines Johnson Health Center Buy Down Recipient FY1996-1997	200 kW PC25	Austin, Texas	Rebekah Baines Johnson Health Center	7/2002 installed	On Going	Natural Gas		Electricity is fed into the Austin Energy electric grid, making it the 1st fuel cell in Texas to feed power to the grid. The health center is using the 900,000 BTUs of thermal energy to heat their water. Austin Energy plans to provide tours and educational programs

UTC Fuel Cells	Logan Energy Buy Down Recipient FY1996-1997	200 kW PAFC	Irvine, CA	Ford Motor Co.'s North American Premier Automotive Group HQ	2001	On going	Natural Gas		
UTC Fuel Cells	Logan Energy Buy Down Recipient FY1996-1997	200 kW PAFC		Merck Co.					
UTC Fuel Cells	Logan Energy Buy Down Recipient FY1996-1997	200 kW PAFC	Bellarie (Houston), TX	ChevronTexaco Company	5/2000				
UTC Fuel Cells	Logan Energy Buy Down Recipient FY1996-1997	200 kW PAFC	Harvey, Louisiana	Citizens Utilities' operations center		Decommissioned	Natural Gas		Project decommissioned after customer relocated. Power plant is available for sale or lease to new customer site ($400,000). See Logan Energy: http://www.loganenergy.com /commercial/saleorlease.html
UTC Fuel Cells	Lord & Company Buy Down Recipient FY1998	Two 200 kW PAFC systems							
UTC Fuel Cells	Lord & Company Buy Down Recipient again FY1998	200 kW PAFC							

UTC Fuel Cells	Los Angeles Department of Water and Power Buy Down recipient FY2000	200 kW PAFC	Near downtown LA on Main Street	LADWP Main Street Service Testing Facility	July 2003	Moved to Storage	Natural Gas	Kjell Ostensen 213.367.0011 kjell.ostensen@ladwp.com Bill Glauz, Manager of Distributed Generation Los Angeles Department of Water and Power; Phone = 213 267-0410; Fax = 213 367-0777; william.glauz@ladwp.com
UTC Fuel Cells	McBride Energy Buy Down recipient FY1998	Two 200 kW PAFC	New Jersey	Ramapo College of New Jersey			Natural Gas	
UTC Fuel Cells	NASA		Space		1958			Since 1966, all of the more than 100 U.S. manned space flights, including the Space Shuttle, have operated with fuel cells supplied by UTC companies. Between their first flight in 1981 and their 71st flight in 1995, the UTC Fuel Cells power plants aboard the Space Shuttle vehicles accumulated a total of 50,000 hours of mission time in 1995
UTC Fuel Cells	National Rural Electric Cooperative Buy Down Recipient FY1995	200 kW PAFC						
UTC Fuel Cells	New England Power Buy Down Recipient FY1995	200 kW PAFC						

UTC Fuel Cells	Co-funded by KeySpan Corp. and DOE	Two 200 kW PC25 Model C systems	Staten Island, NY	Sun Chemical Corp.	mid-1996	Operational	Natural Gas		These units have have operated over 40,000 hours (US record)
UTC Fuel Cells	New York Power Authority with KeySpan	PC25 200 kW	Coney Island, NY	20% of New York Aquarium's power needs	Sept. 2001	Operational	Natural Gas		First Model C's installed in North America Keyspan is an authorized service provider for nine PC25 natural gas fueled fuel cells in the New York area (6/02) Grid Parallel
UTC Fuel Cells	NYPA, KeySpan, and Central Park Police Precinct	200 kW	Manhattan, NY	Central Park HQs	April 1999	On going	Natural Gas		Independent of the electric gird
UTC Fuel Cells	NYPA, KeySpan, And North Central Bronx Hospital (NCBH)	200 kW	Bronx, NY	North Central Hospital	Oct. 2000	On going	Natural Gas		Supplies supplemental power and back-up power
UTC Fuel Cells	NYPA, KeySpan With Westchester County Wastewater Treatment Plant in Yonkers, NY	200 kW	Yonkers, NY	Wastewater treatment plant	April 1997	On going	Anaerobic digester gas from the waste water treatment process		Supplies supplemental power Grid Parallel **World's First ADG fueled fuel cell**
UTC Fuel Cells	NYPA, DEP	2 200kW	Brooklyn, NY	Ward 26 Wastewater Treatment Plant	Feb 2004	Installed			Uses fuel cells to convert waste gas into energy to power the facility Edward Skylay, Jordan Barawitz 212-788-2958 Charles Stucken 718-595-6600

UTC Fuel Cells	NYPA and NYSERDA	3 PC25 200 kW	Coster Street and Rywana Ave. Bronx, NY 10474	Hunts Point waste water treatment plant	2004	On Going	Digester Gas	Grid Parallel
UTC Fuel Cells	NYPA and NYSERDA	2 PC25 200 kW	63 Flushing Ave. Brooklyn, NY 11205	Red Hook Waste Water Treatment Plant	2004	On Going	Digester Gas	Grid Parallel
UTC Fuel Cells	NYPA and NYSERDA	2 PC25 200 kW	43-01 Berrian Boulevard, Astoria NY 11105	Bowery Bay Waste Water Treatment Plant	2002	On Going	Digester Gas	Grid Parallel
UTC Fuel Cells	NYPA and NYSERDA	1 PC25 200 kW	751 Mill Road, Staten Island, NY 10306	Oakwood Beach Waste Water Treatment Plant	2004	On Going	Digester Gas	Grid Parallel
UTC Fuel Cells	Chemical Plant in Staten Island, NY	2 PC25 200 kW	Staten Island, NY	Chemical Plant		On Going	Natural Gas	> 40,000 hours logged
UTC Fuel Cells	Niagara Mohawk aka Plum Street Enterprises Buy Down recipient again in FY1996-1997	200 kW PAFC		Liverpool High School			Natural Gas	
UTC Fuel Cells	NORAM aka Reliant Energy Co. with Logan Energy Buy Down Recipient FY1996-1997	200 kW PAFC	Gulfport, Mississippi	Navy Combat Construction Battalion Base	7/1999	On Going	Natural Gas	
UTC Fuel Cells	Oak Ridge National Laboratory Buy Down Recipient FY2000	200 kW PAFC					Natural Gas	

UTC Fuel Cells	Omaha Public Power District Buy Down Recipient FY1999	200 kW PAFC	Omaha, NE				Natural Gas		
UTC Fuel Cells	Onondaga-Courtland-Madison BOCES Buy Down Recipient in FY1995	200 kW PAFC	4340 WETZEL RD LIVERPOOL, NY 13090	Liverpool High School	Dec. 1999	On Going	Natural Gas		Serves as an educational resource for science teachers and eventually will allow the high school to become an emergency shelter during community disasters.
UTC Fuel Cells	Onondaga-Courtland-Madison BOCES	200 kW PC25 PAFC	Onondaga-Courtland-Madison BOCES, NY	BOCES Regional Information Center (Thompson Road)	1997	On Going	Natural Gas		
UTC Fuel Cells	Sieco SA Buy Down Recipient in FY2000	Three 200 kW PAFC	Brazil, Parana district	Copel Data Center			Natural Gas		COPEL = Companhia Paranaense de Energia (UTC's distribution agent for Latin America)
UTC Fuel Cells	Sieco SA	200 kW PC25 PAFC	Rio de Janeiro, Brazil	Petrobras' R&D Center	Sold in 2002	Being Installed	Natural Gas		Petrobras is Brazil's national oil, gas and energy company. This is the fourth unit to be installed in South Amercia.
UTC Fuel Cells	State of Alaska Buy Down Recipient FY1995	Two 200 kW PAFC	Alaska				Natural Gas		
UTC Fuel Cells	State of Connecticut Buy Down Recipient FY1995	200 kW PAFC	Connecticut				Natural Gas		

Company	Partners	System	Location	Site	Date	Status	Fuel	Image	Notes
UTC Fuel Cells	TBE Germany Buy Down Recipient FY1998	200 kW PAFC systems	Germany	St. Anges Hospital			Natural Gas		Adsorption Chillers are used to have the fuel cell produce power, heat, and air-conditioning (Non-Nuclear Energy – Thermie program) Dr.-Ing. W. Kuhne DBI Gas +49/3731/365 253
UTC Fuel Cells	DBI Gas (Germany), Technische Universitat Dresden, Gastec N.V.	200 kW PAFC system PC25 C	Kamenz, Germany	Maltesser Hospital	Feb. 2000 (w/ AC – May 2000)	37 month project	Natural Gas		
UTC Fuel Cells	Tokyo Gas and Toshiba Corporation	UTC 200 kW PC25A	Tokyo, Japan	Tokyo Gas Research & Development facility	April 13, 1992	4/13/92 - 2000			
UTC Fuel Cells	Tokyo Electric Power Co. (TEPCO) and Toshiba Corporation	11 MW PAFC unit	Tokyo, Japan	Goi Station	March 7 1991	Completed March 1997			This unit accumulated 23,140 hours by 10/00
UTC Fuel Cells	Tokyo Electric Power Co. (TEPCO)	200 kW PC25 PAFC	Japan	Shibaura DHC	March 13 1989	Completed March 1997			This unit accumulated 45,333 hours by 10/00
UTC Fuel Cells	Mitsubishi Electric and Toshiba Corporation	200 kW PAFC PC25	Japan	Kyobashi DHC	Feb. 14 1994	Completed Oct. 2000			This unit accumulated 43,139 hours by 10/00
UTC Fuel Cells	Tokyo Electric Power Co. (TEPCO) and Toshiba Corporation	200 kW PC25 PAFC	Japan	TEPCO R&D Centers	Sept. 20 1994	On going	City gas		This unit accumulated 44,011 hours by 10/00

Manufacturer	Customer/Owner	Unit	Location	Site	Installed	End of Life	Fuel	Image	Notes
UTC Fuel Cells	Toshiba Corporation Southern California Gas Buy Down Recipient FY1995	200 kW PC25 Model A	Anaheim, CA	Kaiser Hospital	May 1993	End of Life May 2000	Natural Gas		Charles Butler: (213) 803-7432 Note: All End of Life units are removed and placed in storage waiting dismantling and disposal
UTC Fuel Cells	Kaiser Permanente HMO Group, Southern California Gas Company, the Gas Research Institute, and the DOE	Two 200 kW PC25 Model A	Riverside, CA	Riverside Medical Center (Kaiser Hospital)	Sept. 1994	Feb. 2000 + March 2001	Natural Gas		Charles Butler: (213) 803-7432 Won the 1994 Efficient Building Award for Energy and the Environment sponsored by Energy User News
UTC Fuel Cells	Toshiba Corporation Buy Down Recipient again FY1995	200 kW PAFC	Heag, Germany				Natural Gas		
UTC Fuel Cells	Toshiba Corporation Buy Down Recipient again FY1995	200 kW PAFC	Japan	Osaka Gas FC R&D Center			Natural Gas		
UTC Fuel Cells	Toshiba Corporation-Power Systems and Service Co. Buy Down Recipient FY1998	200 kW PAFC	Japan	Toshiba Fuchu Works			Natural Gas		

Company	Recipient / Project	Type	Location	Application	Date	Status	Fuel	Image	Notes
UTC Fuel Cells	Washington Water and Power, Buy Down Recipient FY1995	200 kW PAFC	Washington	Avista / Double Tree Inn			Natural Gas		
UTC Fuel Cells	Woking Borough Council, Buy Down Recipient FY1998	200 kW PAFC	United Kingdom	Recreation Center at a local pool			Natural Gas		During the winter, it provides heat and electricity; during the summer it powers the air conditioning units
UTC Fuel Cells	Yankee Gas Services, Buy Down Recipient FY1996-1997	Two 200 kW PAFC systems	Meriden, CT	Yankee Gas Services Office	Oct. 1997		Natural Gas		
UTC Fuel Cells	Verizon, Long Island, NY, Buy Down Recipient FY1999 ($1,400,000)	Seven 200 kW PAFC systems	741 Zeckendorf Blvd. Garden City, Long Island, NY	Major call-routing center serving 40,000 phone lines. Call switching center	2004		Natural Gas		This will be the largest fuel cell deployment project in the world. Mark Marchand, Verizon: 518-396-1080, mark.a.marchand@verizon.com
UTC Fuel Cells	Hotel in Ulsan, South Korea	200 kW PC25 PAFC	Ulsan, South Korea	Hotel			Natural Gas		
UTC Fuel Cells	Sapporo Brewery (AGD) in Chiba, Japan	200 kW PC25 PAFC	Chiba, Japan	Sapporo Brewery		On Going	Digester Gas		
UTC Fuel Cells	Asahi Brewery	200 kW PC25 PAFC	Shikoku Brewery, Japan	Asahi Brewery		On Going	Digester Gas		

Company	Description	Size / Type	Location	Application	Date	Status	Fuel	Image	Notes
UTC Fuel Cells	School in Kaltenkirschen, Germany		Kaltenkirshen Germany	School			Natural Gas		Since the completion of 42 GRI field test sites in the US and Japan, gas utilities have purchased about 22 200 kW PC25s, four of which are still operating
UTC Fuel Cells	Virginia Power and Virginia Natural Gas in a Gas Research Institute (GRI) field test project	40 kW PAFC cogeneration unit	Old Dominium University in Norflok, Virginia	600 bed dormitory	Stalled in 1986	Concluded	Natural Gas		Charles Butler: (213) 803-7432
UTC Fuel Cells	Southern California Gas Company (SoCAL Gas)	200 kW PAFC PC25 Model A	Santa Barbara, CA	Electricity & hot water to Santa Barbara jail	October 1994	Life Ended March 2001	Natural Gas		Charles Butler: (213) 803-7432 Note: All End of Life units are removed and placed in storage waiting dismantling and disposal
UTC Fuel Cells	Southern California Gas	200 kW PC25 Model A	Irvine, CA	Hyatt Hotel	Sept. 1992	Life Ended March 2002	Natural Gas		Charles Butler: (213) 803-7432 Note: All End of Life units are removed and placed in storage waiting dismantling and disposal
UTC Fuel Cells	Southern California Gas	200 kW PC25 Model A	Buena Park, CA	Kraft Foods	July 1993	Removed June 1996	Natural Gas	N/a	Charles Butler: (213) 803-7432 Sold to City of Mesa, AZ
UTC Fuel Cells	Southern California Gas	200 kW PC25 Model A	Los Angeles, CA	SCAQMD Office Building	April 1992	On Going	Natural Gas		Charles Butler: (213) 803-7432 New Cell Stack Installed in Aug. 2001

								Contact
UTC Fuel Cells	Southern California Gas	200 kW PC25 Model A	Santa Barbara, CA	University of California	Sept. 1993	Life Ended June 1998	Natural Gas	Charles Butler: (213) 803-7432
UTC Fuel Cells	Commonwealth Gas	PC25 demonstration project	Fall River, MA	US Army Natick Laboratories			Natural Gas	Note: All End of Life units are removed and placed in storage waiting dismantling and disposal — Peter McGrath: (508) 481-7900
UTC Fuel Cells	Equitable Resources	200 kW PAFC PC25	Pittsburgh, PA	Presbyterian Nursing Home			Natural Gas	Keith Spitznagel: (412)261-3000
UTC Fuel Cells	Equitable Resources	200 kW PAFC PC25	Pittsburgh, PA	Riverside Nursing Home			Natural Gas	Keith Spitznagel: (412)261-3000
UTC Fuel Cells	Peoples' Gas & Light	200 kW PAFC PC25	Chicago, IL	Div. of Street & Meter Repair			Natural Gas	Andrew Plonka: (312) 240-7000
UTC Fuel Cells	Minnegasco	200 kW PAFC PC25	Minneapolis, MN	USAF Reserve Center			Natural Gas	James Radford: (612) 321-4337
UTC Fuel Cells	Jersey Central Power & Light/ GPU	200 kW PAFC PC25	Morristown, NJ	AT&T Research Laboratory			Natural Gas	Steven B. Sanders (201) 455-8328
UTC Fuel Cells	Gas Company of New Mexico	200 kW PAFC PC25	Alberquerque, NM	Kirkland Air Force Base			Natural Gas	Steve Casey: (505) 241-4460
UTC Fuel Cells	Rocherster Gas & Electric	200 kW PAFC PC25	Rochester, NY	Rochester Institute of Technology			Natural Gas	Dan Rider: (716) 724-8322
UTC Fuel Cells	National Fuel Gas	200 kW PAFC PC25	Buffalo, NY	Riefler Concrete			Natural Gas	Betsy Herzog (716) 857-7890

Company	Owner / Program	Size & Type	Location	Facility	Start date	Status	Fuel	Image	Notes
UTC Fuel Cells	AB Parking Facilities, LLC Logan Energy Buy Down Recipient FY1996-1997	3 200 kW PAFC PC25s	Fresno, CA	AB Parking Facilities	Start date 5/2002		Natural Gas		Absorption chiller with fuel cells' waste heat will be used to provide 100 tons of cooling to 12 story commercial building
UTC Fuel Cells	Omaha Zoo	200 kW PAFC PC25	Omaha, Nebraska	Lied Jungle @ Omaha Zoo	Installed 2002	On Going	Natural Gas		The PC25's waste heat is used to warm the water in a number of ponds and heat 5,000 gallons of water used for irrigation each night.
UTC Fuel Cells		200 kW PAFC PC25	Boston, MA	Deer Island	Installed	On Going	Natural Gas		
UTC Fuel Cells		200 kW PAFC PC25	Germany	Bocholt Hospital	Installed 2001	On Going	Natural Gas		Has accumulated more than 7,000 hrs. of uninterrupted operation, setting a European record. Apart from the regular replacement of air filters and scheduled maintenance performed by the hospital's own staff, no service has been required.
UTC Fuel Cells	Connecticut Clean Energy Fund (administered by Connecticut Innovations)	200 kW PC25	South Windsor, CT	South Windsor High School	Installed 10/2002	On Going	Natural Gas		First municipal facility to be powered and heated by a fuel cell in CT. This unit will also server as a basis of a comprehensive fuel-cell curriculum.
UTC Fuel Cells	New Jersey Clean Energy Program Installed by Merck & Company	200 kW PC25	Rahway, NJ	Merck & Company's research facility					Installed under the New Jersey Board of Public Utilities (BPU) statewide initiative. They received a $710,000

ZTEK Corp.	Connecticut Clean Energy Fund; The Renewable Resources Group, LLC	25 kW SOFC	Rocky Hill, CT	Dinosaur State Park	Installed	On Going	Natural Gas	rebate under the New Jersey Clean Energy Program and will receive a $200,000 rebate from DOD's "Buy Down" Program
N/a	NJR Power Services Group; Buy Down Recipient FY1999	Twenty 7 kW PEM systems						This system was previously demonstrated at the Tennessee Valley Authority's Huntsville, AL site from 1998 to 2000.
N/a	District Heating System		Toftlund, Denmark	District Heating System				
N/a	FCTEC http://www.fctec.com/ (DOD)		National Energy Technology Laboratory (NETL)		Coming Soon			
N/a	FCTEC http://www.fctec.com/ (DOD)		Air Expeditionary Force Battlelab (AEFB)		Coming Soon			

N/a	Verizon Boston/Long Island	Smaller fuel cell units	Boston, MA and Long Island, NY	Small field equipment sites in a Boston suburb and Long Island	Operational		
N/a	Verizon NY	Medium Sized fuel cell	Up State NY	Medium-sized facility	Being Planned		
N/a	Pyramid Cos., developers of the proposed $2.2 billion Destiny USA mega mall project	As many as 60 200 kW sized units			Being Planned		Wants to be the world's largest "green" building.
N/a	Celanese Ventures	1 fuel cell system (PEMFC)	Frankfurt, Germany	New PEM membrane manufacturing line	8/2002 Installed		Celanese Ventures makes a membrane for PEMFCs which can be operated at the higher temperature of 150 degrees C. Their new production line is powered by a fuel cell system (Perhaps Ballard?)
Fuel-Cell Commercialization Conference of Japan (FCCJ) – a Japanese 2002 summary	Many manufacturers	Various	Japan		12 months demo. nominal	Natural Gas; LPG; Naptha	■ 1 KW unit for residential use ■ 5kW unit for business use ■ 12 units (11 1kW; 1 5kW) in FY2002 ■ Sites spread over Japan ■ Units integrated into existing power grid

appendix

Hydrogen Cost
Calculations

Information adapted from National Research Council, The Hydrogen Economy:
Opportunities, Costs, Barriers, and R&D Needs. The National Academies Press, 2004
(394 pages). ISBN 0-309-09163-2, www.nap.edu

TABLE E–1. COSTS AND CARBON DIOXIDE EMISSIONS:
CENTRAL STATION NATURAL GAS STEAM REFORMING PLANTS

	Current technology		Future technology	
	without sequestration	with sequestration	without sequestration	with sequestration
Capital investment, 10^6 $				
H_2 production	453.39	623.75	326.85	425.54
Distribution	724.75	724.75	532.98	532.98
Dispensing	714.51	714.51	507.42	507.42
Total capital investment	1,892.66	2,063.01	1,367.25	1,465.95
Production costs, $/kg H_2				
Variable costs				
Feed	0.75	0.79	0.71	0.73
Electricity	0.03	0.08	0.03	0.06
Non-fuel O&M[a]	0.01	0.02	0.01	0.01
Subtotal variable costs	0.79	0.89	0.74	0.80
Fixed costs[b]	0.06	0.08	0.04	0.05
Capital charges[c]	0.18	0.25	0.13	0.17
Total production costs	1.03	1.22	0.92	1.02
Distribution costs, $/kg H_2				
Variable costs				
Electricity	0.01	0.01	0.01	0.01
Non-fuel O&M[a]	0.02	0.02	0.01	0.01
Subtotal variable costs	0.03	0.03	0.02	0.02
Fixed costs[b]	0.09	0.09	0.07	0.07
Capital charges[c]	0.29	0.29	0.21	0.21
Total distribution costs	0.42	0.42	0.31	0.31
Dispensing costs, $/kg H_2				
Variable costs				
Electricity	0.14	0.14	0.11	0.11
Non-fuel O&M[a]	0.09	0.09	0.06	0.06
Subtotal variable costs	0.23	0.23	0.17	0.17
Fixed costs[b]	0.05	0.05	0.04	0.04
Capital charges[c]	0.25	0.25	0.18	0.18
Total dispensing costs	0.54	0.54	0.39	0.39

Hydrogen costs, $/kg				
Production	1.03	1.22	0.92	1.02
Distribution	0.42	0.42	0.31	0.31
Dispensing	0.54	0.54	0.39	0.39
CO_2 disposal	--	0.09	--	0.08
Carbon tax	0.13	0.02	0.12	0.02
Total H_2 costs	2.11	2.28	1.73	1.82
Carbon dioxide vented				
kg C/kg H_2	2.51	0.42	2.39	0.35
Direct use	2.45	0.26	2.34	0.24
Indirect use	0.06	0.16	0.05	0.12
kg CO_2/kg H_2	9.22	1.53	8.75	1.30
Direct use	8.99	0.95	8.56	0.88
Indirect use	0.23	0.58	0.18	0.42
Carbon charge				
$/kg	0.13	0.02	0.12	0.02
Carbon sequestered				
kg CO_2/kg H_2	--	8.56	--	7.91
Carbon	--	2.34	--	2.16

[a] Non-fuel O&M = 1%/yr of capital
[b] Fixed costs = 5%/yr of capital
[c] Capital charges = 18%/yr of capital
Source: National Research Council, *The Hydrogen Economy: Opportunities, Costs, Barriers, and R&D Needs*, Appendix E

TABLE E–2. COSTS AND CARBON DIOXIDE EMISSIONS:
CENTRAL STATION COAL GASIFICATION PLANTS

	Current technology		Future technology	
	without sequestration	**with** sequestration	**without** sequestration	**with** sequestration
Capital investment, 10^6 $				
H_2 production	1,151.92	1,177.33	868.18	889.97
Distribution	724.75	724.75	532.98	532.98
Dispensing	714.51	714.51	507.42	507.42
Total capital investment	2,591.18	2,616.59	1,908.58	1,930.38
Production costs, $/kg H_2				
Variable costs				
Feed	0.21	0.21	0.19	0.19
Electricity	0.11	0.17	0.04	0.08
Non-fuel O&M[a]	0.03	0.03	0.02	0.02
Subtotal variable costs	0.35	0.41	0.25	0.30
Fixed costs[b]	0.15	0.15	0.11	0.11
Capital charges[c]	0.46	0.47	0.35	0.36
Total production costs	0.96	1.03	0.71	0.77
Distribution costs, $/kg H_2				
Variable costs				
Electricity	0.01	0.01	0.01	0.01
Non-fuel O&M[a]	0.02	0.02	0.01	0.01
Subtotal variable costs	0.03	0.03	0.02	0.02
Fixed costs[b]	0.09	0.09	0.07	0.07
Capital charges[c]	0.29	0.29	0.21	0.21
Total distribution costs	0.42	0.42	0.31	0.31
Dispensing costs, $/kg H_2				
Variable costs				
Electricity	0.14	0.14	0.11	0.11
Non-fuel O&M[a]	0.09	0.09	0.06	0.06
Subtotal variable costs	0.23	0.23	0.17	0.17
Fixed costs[b]	0.05	0.05	0.04	0.04
Capital charges[c]	0.25	0.25	0.18	0.18
Total dispensing costs	0.54	0.54	0.39	0.39

Hydrogen costs, $/kg				
Production	0.96	1.03	0.71	0.77
Distribution	0.42	0.42	0.31	0.31
Dispensing	0.54	0.54	0.39	0.39
CO_2 disposal	--	0.16	--	0.15
Carbon tax	0.26	0.04	0.23	0.03
Total H_2 costs	2.17	2.19	1.63	1.64
Carbon dioxide vented				
kg C/kg H_2	5.12	0.82	4.56	0.60
Direct use	4.90	0.49	4.50	0.45
Indirect use	0.21	0.33	0.07	0.15
kg CO_2/kg H_2	18.76	3.00	16.73	2.21
Direct use	17.98	1.80	16.49	1.65
Indirect use	0.77	1.20	0.25	0.56
Carbon charge				
$/kg	0.26	0.04	0.23	0.03
Carbon sequestered				
kg CO_2/kg H_2	--	16.19	--	14.84
Carbon	--	4.41	--	4.05

[a] Non-fuel O&M = 1%/yr of capital
[b] Fixed costs = 5%/yr of capital
[c] Capital charges = 18%/yr of capital
Source: National Research Council, *The Hydrogen Economy: Opportunities, Costs, Barriers, and R&D Needs*, Appendix E

TABLE E–3. COSTS OF FUTURE CENTRAL STATION
NUCLEAR THERMAL WATER-SPLITTING PLANTS

	Future technology
Capital investment, 10^6 \$	
H_2 production	2,468.19
Distribution	532.98
Dispensing	507.42
Total capital investment	3,508.59
Production costs, \$/kg H_2	
Variable costs	
Feed	0.20
Decommission fund	0.06
Non-fuel O&M[a]	0.06
Subtotal variable costs	0.32
Fixed costs[b]	0.31
Capital charges[c]	1.00
Total production costs	1.63
Distribution costs, \$/kg H_2	
Variable costs	
Electricity	0.01
Non-fuel O&M[a]	0.01
Subtotal variable costs	0.02
Fixed costs[b]	0.07
Capital charges[c]	0.21
Total distribution costs	0.31
Dispensing costs, \$/kg H_2	
Variable costs	
Electricity	0.11
Non-fuel O&M[a]	0.06
Subtotal variable costs	0.17
Fixed costs[b]	0.04
Capital charges[c]	0.18
Total dispensing costs	0.39
H_2 costs, \$/kg	
Production	1.63
Distribution	0.31
Dispensing	0.39
Total H_2 costs	2.33

[a] Non-fuel O&M = 1%/yr of capital
[b] Fixed costs = 5%/yr of capital
[c] Capital charges = 18%/yr of capital
Source: National Research Council, *The Hydrogen Economy: Opportunities, Costs, Barriers, and R&D Needs*, Appendix E

TABLE E−4. COSTS OF MID-SIZE NATURAL GAS STEAM REFORMING PLANTS

	Current technology		Future technology	
	without sequestration	with sequestration	without sequestration	with sequestration
Capital investment, 10⁶ $				
H₂ production	21.52	29.26	17.12	23.06
Distribution	34.73	34.73	18.54	18.54
Dispensing	20.70	20.70	9.01	9.01
Total capital investment	76.95	84.69	44.67	50.61
Production costs, $/kg H₂				
Variable costs				
Feed	0.79	0.82	0.74	0.79
Electricity	0.04	0.10	0.03	0.08
Non-fuel O&Mᵃ	0.03	0.03	0.02	0.03
Subtotal variable costs	0.86	0.96	0.79	0.90
Fixed costsᵇ	0.13	0.17	0.10	0.13
Capital chargesᶜ	0.40	0.54	0.32	0.43
Total production costs	1.38	1.67	1.21	1.46
Distribution costs, $/kg H₂				
Variable costs				
Labor	0.09	0.09	0.09	0.09
Fuel	0.01	0.01	0.01	0.01
Electricity	0.50	0.50	0.36	0.36
Non-fuel O&Mᵃ	0.06	0.06	0.03	0.03
Subtotal variable costs	0.65	0.65	0.49	0.49
Fixed costsᵇ	0.28	0.28	0.15	0.15
Capital chargesᶜ	0.88	0.88	0.47	0.47
Total distribution costs	1.80	1.80	1.10	1.10
Dispensing costs, $/kg H₂				
Variable costs				
Electricity	0.06	0.06	0.06	0.06
Non-fuel O&Mᵃ	0.13	0.13	0.06	0.06
Subtotal variable costs	0.18	0.18	0.11	0.11
Fixed costsᵇ	0.08	0.08	0.03	0.03
Capital chargesᶜ	0.36	0.36	0.16	0.16
Total dispensing costs	0.62	0.62	0.30	0.30
Hydrogen costs, $/kg				
Production	1.38	1.67	1.21	1.46
Distribution	1.80	1.80	1.10	1.10
Dispensing	0.62	0.62	0.30	0.30
CO₂ disposal	--	0.09	--	0.09
Carbon tax	0.13	0.02	0.12	0.02
Total H₂ costs	3.94	4.20	2.74	2.97

ᵃ Non-fuel O&M = 1%/yr of capital
ᵇ Fixed costs = 5%/yr of capital
ᶜ Capital charges = 18%/yr of capital
Source: National Research Council, *The Hydrogen Economy: Opportunities, Costs, Barriers, and R&D Needs*, Appendix E

TABLE E-5. COSTS OF MID-SIZE BIOMASS GASIFICATION PLANTS

	Current technology		Future technology	
	without sequestration	with sequestration	without sequestration	with sequestration
Capital investment, 10^6 $				
H$_2$ production	121.04	123.91	59.04	60.53
Distribution	34.73	34.73	18.54	18.54
Dispensing	20.70	20.70	9.01	9.01
Total capital investment	176.46	179.33	86.59	88.08
Production costs, $/kg H$_2$				
Variable costs				
Feed	0.98	0.98	0.42	0.42
Electricity	0.29	0.41	0.14	0.21
Non-fuel O&M[a]	0.15	0.16	0.07	0.08
Subtotal variable costs	1.42	1.54	0.64	0.71
Fixed costs[b]	0.77	0.79	0.37	0.38
Capital charges[c]	2.44	2.50	1.19	1.22
Total production costs	4.63	4.82	2.21	2.32
Distribution costs, $/kg H$_2$				
Variable costs				
Labor	0.09	0.09	0.09	0.09
Fuel	0.01	0.01	0.01	0.01
Electricity	0.50	0.50	0.36	0.36
Non-fuel O&M[a]	0.06	0.06	0.03	0.03
Subtotal variable costs	0.65	0.65	0.49	0.49
Fixed costs[b]	0.28	0.28	0.15	0.15
Capital charges[c]	0.88	0.88	0.47	0.47
Total distribution costs	1.80	1.80	1.10	1.10
Dispensing costs, $/kg H$_2$				
Variable costs				
Electricity	0.06	0.06	0.06	0.06
Non-fuel O&M[a]	0.13	0.13	0.06	0.06
Subtotal variable costs	0.18	0.18	0.11	0.11
Fixed costs[b]	0.08	0.08	0.03	0.03
Capital charges[c]	0.36	0.36	0.16	0.16
Total dispensing costs	0.62	0.62	0.30	0.30
Hydrogen costs, $/kg				
Production	4.63	4.82	2.21	2.32
Distribution	1.80	1.80	1.10	1.10
Dispensing	0.62	0.62	0.30	0.30
CO$_2$ disposal	--	0.26	--	0.17
Carbon tax	0.03	−0.31	0.01	−0.21
Total H$_2$ costs	7.07	7.19	3.63	3.68

[a] Non-fuel O&M = 1%/yr of capital
[b] Fixed costs = 5%/yr of capital
[c] Capital charges = 18%/yr of capital
Source: National Research Council, *The Hydrogen Economy: Opportunities, Costs, Barriers, and R&D Needs*, Appendix E

TABLE E–6. COSTS OF MID-SIZE GRID-BASED ELECTROLYSIS PLANTS

	Current technology	Future technology
Capital investment, 10^6 \$		
H_2 production	84.78	9.71
Distribution	34.73	18.54
Dispensing	20.70	9.01
Total capital investment	140.20	37.26
Production costs, \$/kg H_2		
Variable costs		
Feed	−0.08	−0.08
Electricity	2.43	2.11
Non-fuel O&M[a]	0.11	0.01
Subtotal variable costs	2.46	2.04
Fixed costs[b]	0.54	0.06
Capital charges[c]	1.71	0.20
Total production costs	4.70	2.30
Distribution costs, \$/kg H_2		
Variable costs		
Labor	0.09	0.09
Fuel	0.01	0.01
Electricity	0.50	0.36
Non-fuel O&M[a]	0.06	0.03
Subtotal variable costs	0.65	0.49
Fixed costs[b]	0.28	0.15
Capital charges[c]	0.88	0.47
Total distribution costs	1.80	1.10
Dispensing costs, \$/kg H_2		
Variable costs		
Electricity	0.06	0.06
Non-fuel O&M[a]	0.13	0.06
Subtotal variable costs	0.18	0.11
Fixed costs[b]	0.08	0.03
Capital charges[c]	0.36	0.16
Total dispensing costs	0.62	0.30
Hydrogen costs, \$/kg		
Production	4.70	2.30
Distribution	1.80	1.10
Dispensing	0.62	0.30
Carbon tax	0.24	0.20
Total H_2 costs	7.36	3.91

[a] Non-fuel O&M = 1%/yr of capital
[b] Fixed costs = 5%/yr of capital
[c] Capital charges = 18%/yr of capital
Source: National Research Council, *The Hydrogen Economy: Opportunities, Costs, Barriers, and R&D Needs*, Appendix E

TABLE E−7. COSTS AND CARBON EMISSIONS:
ONSITE NATURAL GAS STEAM REFORMING UNITS

	Current technology	Future technology
Capital investment		
10⁶ $	1.85	0.96
H₂ costs, $/kg		
Variable production costs		
Feed	1.37	1.17
Electricity	0.15	0.12
Non-fuel O&M[a]	0.12	0.06
Subtotal variable costs	1.64	1.35
Fixed production costs[b]	0.23	0.12
Capital charges[c]	1.64	0.85
Total production costs	3.51	2.33
Carbon tax	0.17	0.14
Total H₂ costs	3.68	2.47
Carbon dioxide vented		
kg C/kg H₂	3.31	2.82
Direct use	3.11	2.67
Indirect use	0.19	0.15
kg CO₂/kg H₂	12.13	10.34
Direct use	11.42	9.79
Indirect use	0.71	0.55

[a] Non-fuel O&M = 1%/yr of capital
[b] Fixed costs = 5%/yr of capital
[c] Capital charges = 18%/yr of capital
Source: National Research Council, *The Hydrogen Economy: Opportunities, Costs, Barriers, and R&D Needs*, Appendix E

TABLE E–8. COSTS AND CARBON EMISSIONS:
ONSITE ELECTROLYSIS UNITS USING GRID POWER

	Current technology	Future technology
Capital investment		
10^6 $	2.54	0.57
H$_2$ costs, $/kg		
Variable production costs		
Electricity	3.84	3.31
Non-fuel O&M[a]	0.16	0.04
Subtotal variable costs	4.00	3.35
Fixed production costs[b]	0.32	0.07
Capital charges[c]	2.26	0.51
Total production costs	6.58	3.93
Carbon tax	0.24	0.21
Total H$_2$ costs	6.82	4.13
Carbon dioxide vented		
kg C/kg H$_2$	4.79	4.13
Indirect use	4.79	4.13
kg CO$_2$/kg H$_2$	15.13	15.13
Indirect use	15.13	15.13

[a] Non-fuel O&M = 1%/yr of capital
[b] Fixed costs = 5%/yr of capital
[c] Capital charges = 18%/yr of capital
Source: National Research Council, *The Hydrogen Economy: Opportunities, Costs, Barriers, and R&D Needs,* Appendix E

TABLE E–9. COSTS OF ONSITE ELECTROLYSIS UNITS BASED ON WIND TURBINE POWER

	Current technology	Future technology
Capital investment		
10^6 $	6.86	0.89
H$_2$ costs, $/kg		
Variable production costs		
Electricity	3.29	1.90
Non-fuel O&M[a]	0.44	0.06
Subtotal variable costs	3.73	1.95
Fixed production costs[b]	0.87	0.11
Capital charges[c]	6.09	0.79
Total production costs	10.69	2.86
Total H$_2$ costs	10.69	2.86

[a] Non-fuel O&M = 1%/yr of capital
[b] Fixed costs = 5%/yr of capital
[c] Capital charges = 18%/yr of capital
Source: National Research Council, *The Hydrogen Economy: Opportunities, Costs, Barriers, and R&D Needs*, Appendix E

TABLE E–10. COSTS AND CARBON EMISSIONS:
ONSITE ELECTROLYSIS UNITS BASED ON HYBRID WIND-TURBINE/GRID POWER

	Current technology	Future technology
Capital investment		
10^6 $	2.75	0.59
H$_2$ costs, $/kg		
Variable production costs		
Electricity	3.67	2.75
Non-fuel O&M[a]	0.17	0.04
Subtotal variable costs	3.85	2.78
Fixed production costs[b]	0.35	0.07
Capital charges[c]	2.44	0.52
Total production costs	6.64	3.38
Carbon tax	0.17	0.12
Total H$_2$ costs	6.81	3.50
Carbon dioxide vented		
kg C/kg H$_2$	3.35	2.48
Indirect use	3.35	2.48
kg CO$_2$/kg H$_2$	12.28	9.08
Indirect use	12.28	9.08

[a] Non-fuel O&M = 1%/yr of capital
[b] Fixed costs = 5%/yr of capital
[c] Capital charges = 18%/yr of capital
Source: National Research Council, *The Hydrogen Economy: Opportunities, Costs, Barriers, and R&D Needs*, Appendix E

TABLE E–11. COSTS OF ONSITE ELECTROLYSIS UNITS BASED ON SOLAR PV POWER

	Current technology	Future technology
Capital investment		
10^6 $	9.94	1.43
H$_2$ costs, $/kg		
Variable production costs		
Electricity	17.48	4.64
Non-fuel O&M[a]	0.63	0.09
Subtotal variable costs	18.11	4.73
Fixed production costs[b]	1.26	0.18
Capital charges[c]	8.82	1.27
Total production costs	28.19	6.18
Total H$_2$ costs	28.19	6.18

[a] Non-fuel O&M = 1%/yr of capital
[b] Fixed costs = 5%/yr of capital
[c] Capital charges = 18%/yr of capital
Source: National Research Council, *The Hydrogen Economy: Opportunities, Costs, Barriers, and R&D Needs*, Appendix E

TABLE E-12. COSTS AND CARBON EMISSIONS:
ONSITE ELECTROLYSIS UNITS BASED ON HYBRID SOLAR-PV/GRID POWER

	Current technology	Future technology
Capital investment		
10^6 \$	2.74	0.59
H_2 costs, \$/kg		
Variable production costs		
Electricity	6.57	3.58
Non-fuel O&M[a]	0.17	0.04
Subtotal variable costs	6.74	3.61
Fixed production costs[b]	0.35	0.07
Capital charges[c]	2.43	0.52
Total production costs	9.52	4.21
Carbon tax	0.19	0.17
Total H_2 costs	9.71	4.37
Carbon dioxide vented		
kg C/kg H_2	3.83	3.30
Indirect use	3.83	3.30
kg CO_2/kg H_2	14.04	12.11
Indirect use	14.04	12.11

[a] Non-fuel O&M = 1%/yr of capital
[b] Fixed costs = 5%/yr of capital
[c] Capital charges = 18%/yr of capital
Source: National Research Council, *The Hydrogen Economy: Opportunities, Costs, Barriers, and R&D Needs*, Appendix E

TABLE E–13. FUELING STATION COSTS USING CURRENT TECHNOLOGY

Design Basis
2,740 kg/d H_2
4,875 fuel cell vehicles supported
114 kg/h daily average
5 dispensers
22-hour operation

Gaseous hydrogen*		Liquid hydrogen**	
Capital costs, 10^6 $			
Compressors	0.62	Liquid pump/vaporizer	0.506
Buffer storage	0.38	Liquid storage	0.529
Dispensers	0.07	Buffer storage	0.403
Subtotal equipment	1.07	Dispensers	0.075
		Subtotal equipment	1.513
Facilities/permitting	0.27	Facilities/permitting	0.378
Engineering/startup	0.11	Engineering/startup	0.151
Contingencies	0.11	Contingencies	0.151
Working capital/misc.	0.08	Working capital/misc.	0.106
Total capital costs	1.63	Total capital costs	2.300
Hydrogen costs, $/kg H_2			
Variable costs		Variable costs	
Non-fuel O&M	0.09	Non-fuel O&M	0.13
Electricity	0.14	Electricity	0.06
Subtotal variable costs	0.23	Subtotal variable costs	0.18
Fixed operating costs	0.05	Fixed operating costs	0.08
Capital charges	0.25	Capital charges	0.36
Total	0.54	Total	0.62

* Hydrogen delivered by pipeline from central station, one of 438 stations supplied. Pipeline capital costs estimated at $725 million and operating costs at $0.42/kg H_2 using current technology.

** Hydrogen delivered by tanker truck from mid-size plant, one of 9 stations supplied. Trucking capital costs estimated at $34.7 million (including trucks) and operating costs at $1.80/kg H_2 using current technology.

Source: National Research Council, *The Hydrogen Economy: Opportunities, Costs, Barriers, and R&D Needs*, Appendix E

TABLE E−14. FUELING STATION COSTS USING FUTURE TECHNOLOGY

Design Basis
2,740 kg/d H_2
4,875 fuel cell vehicles supported
114 kg/h daily average
5 dispensers
22-hour operation

Gaseous hydrogen*		Liquid hydrogen**	
Capital costs, 10^6 \$			
Compressors	0.31	Liquid pump/vaporizer	0.304
Buffer storage	0.38	Liquid storage	0.113
Dispensers	0.07	Buffer storage	0.202
Subtotal equipment	0.76	Dispensers	0.040
		Subtotal equipment	0.659
Facilities/permitting	0.19	Facilities/permitting	0.165
Engineering/startup	0.08	Engineering/startup	0.066
Contingencies	0.08	Contingencies	0.066
Working capital/misc.	0.05	Working capital/misc.	0.046
Total capital costs	1.16	Total capital costs	1.001
Hydrogen costs, \$/kg H_2			
Variable costs		Variable costs	
Non-fuel O&M	0.06	Non-fuel O&M	0.06
Electricity	0.11	Electricity	0.06
Subtotal variable costs	0.17	Subtotal variable costs	0.11
Fixed operating costs	0.04	Fixed operating costs	0.03
Capital charges	0.18	Capital charges	0.16
Total	0.39	Total	0.30

* Hydrogen delivered by pipeline from central station, one of 438 stations supplied. Pipeline capital costs estimated at $533 million and operating costs at \$0.31/kg H_2 using future technology.

** Hydrogen delivered by tanker truck from mid-size plant, one of 9 stations supplied. Trucking capital costs estimated at \$18.5 million (including trucks) and operating costs at \$1.10/kg H_2 using future technology.

Source: National Research Council, *The Hydrogen Economy: Opportunities, Costs, Barriers, and R&D Needs*, Appendix E

appendix

Gaseous Hydrogen Safety

Information created by Air Products
(www.airproducts.com/NR/rdonlyres/3C6D640E-93C5-4
BD0-8F21-8F7344C66554/0/safetygram4.pdf)

General

Hydrogen is a colorless, odorless, tasteless, highly flammable gas. It is also the lightest-weight gas. Since hydrogen is noncorrosive, special materials of construction are not usually required. However, embrittlement occurs in some metals at elevated temperatures and pressures. Stationary vessels and piping should be designed to the American Society of Mechanical Engineers (ASME) code and the American National Standards Institute (ANSI) Pressure Piping code for the pressures and temperatures involved. Vessels used for transportation must be designed to meet the Department of Transportation (DOT) code.

Gaseous hydrogen may be supplied in tube trailers and cylinders. Hydrogen is usually compressed into gas cylinders by oil-lubricated compressors. The amount of gas in a cylinder is determined by the pressure, temperature, cylinder size, and cylinder pressure rating.

The molecular symbol for hydrogen is H_2.

Health

Hydrogen gas is odorless and nontoxic but may induce suffocation by diluting the concentration of oxygen in air below levels necessary to support life.

Caution: The amount of hydrogen gas necessary to produce oxygen-deficient atmospheres is well within the flammable range, making fire and explosion the primary hazards associated with hydrogen and air atmospheres.

Flammability

The wide flammability range, 4% to 74% in air, and the small amount of energy required for ignition necessitate special handling to prevent the inadvertent mixing of hydrogen with air. Care should be taken to eliminate sources of ignition such as sparks from electrical equipment, static electricity sparks, open flames, or any extremely hot objects.

Hydrogen and air mixtures, within the flammable range, can explode and may burn with a pale blue, almost invisible flame.

Manufacture

Hydrogen is produced by the steam reforming of natural gas, the electrolysis of water, the dissociation of ammonia, and as a by-product of petroleum distillation and chlorine manufacture, with the primary method for on-purpose generation being the steam reforming of natural gas. Other feed-stocks can include ethane, propane, butane, and light and heavy naphtha but are not commonly used. The steam reforming process produces syngas, which is a mixture of hydrogen and carbon monoxide. Regardless of the method of production, the product steam is then separated into its components and the hydrogen dried, purified, and compressed into cylinders or tubes for transportation.

Uses

Hydrogen is used in the chemical industry to synthesize ammonia and in the hydrogenation of vegetable and animal oils and fats.

In the metallurgical industry, hydrogen is used to reduce metal oxides and prevent oxidation in heat-treating certain metals and alloys. Some use of hydrogen is made in the welding and cutting of metals. Hydrogen is also used by semiconductor manufacturers, primarily to form reducing atmospheres.

TABLE F-1 PROPERTIES

Molecular Weight	2.016
Boiling Point @ 1 atm	−423.2°F (−252.9°C)
Freezing Point @ 1 atm	−434.8°F (−259.3°C)
Critical Temperature	−400.4°F (−240.2°C)
Critical Pressure	186 psia (12.7 atm)
Density, Liquid @ B.P., 1 atm	4.42 lb./cu.ft.
Density, Gas @ 68°F (20°C), 1 atm	0.005229 lb./cu.ft.
Specific Gravity, Gas (Air = 1) @ 68°F (20°C), 1 atm	0.0696
Specific Gravity, Liquid @ B.P., [water=1 @ 68°F (20°C)]	0.0710
Specific Volume @ 68°F (20°C), 1 atm	191 cu. ft./lb.
Latent Heat of Vaporization	389 Btu/lb. mole
Flammable Limits @ 1 atm in air	4.00%–74.2% (by Volume)
Flammable Limits @ 1 atm in oxygen	3.90%–95.8% (by Volume)
Detonable Limits @ 1 atm in air	18.2%–58.9% (by Volume)
Detonable Limits @ 1 atm in oxygen	15%–90% (by Volume)
Autoignition Temperature @ 1 atm	1060°F (571°C)
Expansion Ratio, Liquid to Gas, B.P. to 68°F (20°C)	1 to 845

Containers

Cylinders

Cylinders are intended to be secured and stored upright. They are tapered to a small opening on the top. The open end is threaded to receive a cylinder valve or other suitable outlet connection. A threaded neckring is secured to the tapered end of the cylinder to allow a protective cap to be installed. Cylinders may be used individually or can be manifolded together to allow for larger gas storage volume.

Tubes

Tubes are generally mounted on truck-trailer chassis, referred to as tube trailers. Stationary tube or hydril tube modules store larger quantities of hydrogen at customer locations. The tubes are tapered on both ends. Each end has threaded openings to which connections, valves, or safety devices can be attached. The amount of hydrogen contained in each tube depends on tube diameter, length, and pressure rating.

Figure F–1 depicts a typical tube trailer for transporting large volumes of gaseous hydrogen. Tube trailers are available in capacities up to 126,000 standard cubic feet and 2,640 psig pressure.

Fig. F–1. Hydrogen Tube Trailer

A typical module bulk gaseous hydrogen system is depicted in Figure F–2. Modules are available in 3 to 18 tube configurations with capacities to 150,000 standard cubic feet of hydrogen. Mobile and stationary tubes have individual valves and safety devices but are manifolded together so that the customer can withdraw product from a single tube or several tubes at a time. Tube modules can be filled to 2,400 psig.

Specifications

Cylinders and mobile tubes are manufactured according to DOT-3A or DOT-3AA specifications. Cylinders and mobile tubes are hydrostatically tested upon manufacture and tested periodically thereafter at 5/3 times the service pressure, as specified by DOT regulations.

Hydrogen may be stored in ASME coded and stamped, National Board registered high-pressure gas storage tubes as part of a stationary installation. These tubes are hydrostatically tested by the manufacturer but, unlike cylinders and mobile tubes, periodical hydrostatic testing is not required.

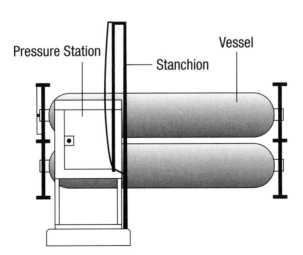

FIG. F–2. A TYPICAL TUBE CONTAINER SYSTEM FOR BULK GAS

Valves

The Compressed Gas Association and the American National Standards Institute have recommended a thread size of 0.825 inch–14 external left-hand threads per inch, designated as valve connection No. 350. Further information on valves is provided in Air Products' Safetygram-23, "Cylinder Valves," and Safetygram-31, "Cylinder Valve Outlet Connections."

Pressure Relief Devices

Pressure relief devices provide protection against excessive pressure in the container. Pressure relief devices are integral parts of the cylinder valves. They are also recommended for use on pressurized systems. These devices take the form of frangible disks or pressure relief valves. Further information on pressure relief devices is provided in Safetygram-15.

Identification: Cylinders and Mobile Tubes

Each cylinder or mobile tube is identified by stampings in the metal of the cylinder shoulder. Figure F–3 depicts an example of these markings and what they mean.

Shipment: Hydrogen Cylinders

The shipment of hydrogen cylinders by surface transportation must conform to DOT regulations as set forth in the Code of Federal Regulations, Title 49, which describes the marking, labeling, placarding, and shipping papers required. A DOT 4"x4" red, flammable label, as illustrated in Figure F–4, is required for common carrier shipments. For emergency response, refer to UN Number 1049 in the Department of Transportation's "Emergency Response Guide." Shipments by air must conform with Title 49 in the Code of Federal Regulations. Final acceptance for air transport is at the discretion of the airline.

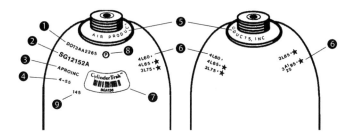

1. Cylinder Specification

   • DOT – Department of Transportation, which is the regulatory body that governs the use of cylinders.

   • Specification of the cylinder type of material of construction (e.g., 3AA).

   • Service or working pressure in pounds per square inch (e.g., 2,265 psi).

2. Cylinder Serial Number

   • The letters SG precede the serial numbers for Specialty Gas cylinders.

3. Registered Owner Symbol

   • Symbol used to indicate the original owner of the cylinders.

   • APROINC is a Registered Owner Symbol for Air Products.

4. Date of Manufacture

   • This date (month-year) also indicates the original hydrostatic test.

5. Neck Ring Identification

   • The cylinder neck ring displays the name of the current owner of the cylinder.

6. Retest Markings

   • The format for a retest marking is: Month-Facility-Year-Plus Rating-Star Stamp.

   • The + symbol (Plus Rating) indicates that the cylinder qualifies for 10% overfill.

   • The ★ symbol (Star Stamp) indicates that the cylinder meets the requirements for 10-year retest.

7. CylinderTrak™ Bar Code Label

   • The CylinderTrak bar code label provides a unique cylinder identifier and is used by computer systems to track cylinders throughout the fill process. As an optional service, we have the capability of tracking cylinders to and from customers.

8. Cylinder Manufacturer's Inspection Marking

9. Cylinder Tare (Empty) Weight

FIG. F–3. KEY TO CYLINDER STAMPINGS

Safety Considerations

The hazards associated with handling hydrogen are fire, explosion, and asphyxiation. The potential for forming and igniting flammable mixtures containing hydrogen may be higher than for other flammable gases because:

1. Hydrogen migrates quickly through small openings.
2. The minimum ignition energy for flammable mixtures containing hydrogen is extremely low.

Burns may result from unknowingly walking into a hydrogen fire. The fire and explosion hazards can be controlled by appropriate design and operating procedures. Preventing the formation of combustible fuel-oxidant mixtures and removing or otherwise inerting potential sources of ignition (electric spark, static electricity, open flames, etc.) in areas where the hydrogen will be used are essential. Careful evacuation and purge operations should be used to prevent the formation of flammable or explosive mixtures. Adequate ventilation will help reduce the possible formation of flammable mixtures in the event of a hydrogen leak and will also eliminate the potential hazard of asphyxiation.

Purging

Gaseous hydrogen systems must be purged of air, oxygen, or other oxidizers prior to admitting hydrogen to the systems, and purged of hydrogen before opening the system to the atmosphere. Purging should be done to prevent the formation of flammable mixtures and can be accomplished in several ways.

Piping systems and vessels intended for gaseous hydrogen service should be inerted by suitable purging or evacuation procedures. If the piping systems are extensive or complicated, successive evacuations broken first by an inert gas and finally with hydrogen are most reliable.

Evacuating and purging of equipment in gaseous hydrogen service should include the following considerations:

1. Evacuate the equipment and break vacuum with an inert gas, such as nitrogen. Purge with an inert gas if equipment design does not permit evacuation.

2. Repeat step 1 at least three times. If analytical equipment is available, purge system until oxygen content of residual gas is either less than or meets the process specification impurity level.

3. Hydrogen may now be introduced to the equipment.

4. Flush system with hydrogen until required purity is reached. Vent all waste hydrogen through a flue or flare stack.

Any purge method should be repeated as often as required to be certain a flammable mixture cannot be formed upon introducing hydrogen or air to the system.

Buildings

1. Provide adequate ventilation, particularly in roof areas where hydrogen might collect. Forced ventilation may be necessary in some applications.

2. The atmosphere in areas in which hydrogen gas may be vented and might collect should be tested with a portable or continuous flammable gas analyzer.

3. Provide an explosion venting surface or vents, taking care to vent a pressure wave to areas where people or other equipment will not become involved. Explosion vents may not be required where small quantities of hydrogen are involved.

4. Buildings should be electrically grounded.

5. Electrical equipment must conform to the existing National Electrical Codes. Electrical equipment not conforming must be located outside the electrical area classified as hazardous. All electrical equipment must be grounded.

6. Building materials should be noncombustible.

7. Post "No Smoking" and "No Open Flames" signs.

Fig. F−4. Flammable Gas Shipping Label

Handling and Storage of Cylinders

1. Never drop cylinders or permit them to strike each other violently.

2. Cylinders should be assigned a definite area of storage. The area should be dry, cool, well-ventilated, and preferably fire-resistant. Keep cylinders protected from excessive temperatures by storing them away from radiators or other sources of heat.

3. Cylinders may be stored in the open, but in such cases should be protected against extremes of weather and from damp ground to prevent rusting.

4. The valve protection cap should be left in place until the cylinder has been secured against a wall, a bench, or placed in a cylinder stand and is ready to be used.

5. Avoid dragging or sliding cylinders, even for short distances. Cylinders should be moved by using a suitable hand truck.

6. Do not use cylinders as rollers for moving material or other equipment.

7. Never tamper with safety devices in valves or cylinders.

8. When returning empty cylinders, close the valve before shipment. Leave some positive pressure in the cylinder. Replace any valve outlet and protective caps originally shipped with the cylinder. Mark and label the cylinder as "Empty." Do not store full and empty cylinders together.

9. No part of a cylinder should be subjected to a temperature above 125°F (52°C). Prevent sparks or flames from welding or cutting torches or any other source coming in contact with cylinders. Do not permit cylinders to come in contact with electrical apparatus or circuits.

10. Never permit oil, grease, or other readily combustible substances to come in contact with cylinders or valves.

11. Use regulators and pressure relief devices when connecting cylinders to circuits having lower pressure service ratings.

12. Smoking or open flames should be prohibited in hydrogen cylinder and tube storage and use areas.

13. Know and understand the properties, uses, and safety precautions of hydrogen before using the gas and associated equipment. Consult the Air Products Material Safety Data Sheet (MSDS) for safety information.

14. Always open a cylinder valve slowly. Never crack open a hydrogen cylinder to clear the valve of dust, as the escaping hydrogen may ignite.

15. Total storage capacity of an indoor hydrogen system should be limited to 3000 cubic feet or less.

16. Hydrogen storage inside a building should not be near oxidants or other combustible storage.

17. When finished with a cylinder, always close the valve. When work is to be interrupted for any length of time, the valve should be closed and all gas released from the hose and regulator to a safe location.

18. If a cylinder or valve is defective or leaking, remove the cylinder to a remote outdoor location away from possible sources of ignition, and post the area as to the hazard involved. Notify your supplier.

19. If a cylinder protective cap is extremely difficult to remove, do not apply excessive force or pry the cap loose with a bar inserted into the ventilation openings. Attach a label or tag to the cylinder identifying the problem and return the cylinder to the supplier.

20. Wrenches should not be used on valves equipped with a hand-wheel. If the valve is faulty, attach a label or tag to the cylinder identifying the problem and return the cylinder to the supplier.

21. Compressed gas cylinders should not be refilled except by qualified producers of compressed gases.

22. Shipment of a compressed gas cylinder filled without the consent of the owner is a violation of Federal Law.

Bulk Gaseous Storage

Location – General Requirements

A. The system should be located so that is readily accessible to delivery equipment and to authorized personnel.

B. Systems must be located above ground.

C. Systems should not be located beneath electric power lines.

D. Systems should not be located close to flammable liquid piping or piping of other flammable gases.

E. It is advisable to locate the system on ground higher than flammable liquid storage or liquid oxygen storage. Where it is necessary to locate the system on ground that is lower than adjacent flammable liquid storage or liquid oxygen storage, suitable protective means (such as by diking, diversion curbs, or grading) should be taken.

F. The hydrogen storage location should be permanently placarded: "Hydrogen – Flammable Gas – No Smoking – No Open Flames," or equivalent.

G. The area within 15 feet of any hydrogen container should be kept free of dry vegetation and combustible material.

TABLE F–2. MINIMUM DISTANCE (FT) FROM GASEOUS HYDROGEN SYSTEMS TO EXPOSURES*

Type of Outdoor Exposure	Size of Hydrogen System		
	Less than 3,500 SCF (99 m³)	3,500 SCF (99 m³) to 15,000 SCF (425 m³)	In Excess of 15,000 SCF (425 m³)
1. Building/Structure			
a) Wall(s) adjacent to system constructed of noncombustible or limited-combustible materials			
1) Sprinklered building/structure or contents noncombustible.	0[1]	5[1]	5[1]
2) Unsprinklered building/structure with combustible contents.			
Adjacent wall(s) with fire-resistance rating less than 2 hours.[2]	0[3]	10	25[4]
Adjacent wall(s) with fire-resistance rating of 2 hours or greater.[2]	0	5	5
b) Wall(s) adjacent to system constructed of other than noncombustible or limited-combustible materials	10	25	50[4]
2. Wall Openings			
a) Not above any part of a system	10 ft (3.1 m)	10 ft (3.1 m)	10 ft (3.1 m)
b) Above any part of a system	25 ft (7.6 m)	25 ft (7.6 m)	25 ft (7.6 m)
3. All Classes of Flammable and Combustible Liquids Above Ground*			
a) 0–1,000 gallons	10 ft (3.1 m)	25 ft (7.6 m)	25 ft (7.6 m)
b) In excess of 1,000 gallons	25 ft (7.6 m)	50 ft (15.2 m)	50 ft (15.2 m)
4. All Classes of Flammable and Combustible Liquids Below Ground— 0–1,000 gallons**			
a) Tank	10 ft (3.1 m)	10 ft (3.1 m)	10 ft (3.1 m)
b) Vent or fill opening of tank	25 ft (7.6 m)	25 ft (7.6 m)	25 ft (7.6 m)

5.	All Classes of Flammable and Combustible Liquids Below Ground— In Excess of 1,000 Gallons**			
	a) Tank	20 ft (6.1 m)	20 ft (6.1 m)	20 ft (6.1 m)
	b) Vent or fill opening of tank	25 ft (7.6 m)	25 ft (7.6 m)	25 ft (7.6 m)
6.	Flammable Gas Storage (Other than Hydrogen), Either High Pressure or Low Pressure			
	a) 0–15,000 CF (255 m³) capacity	10 ft (3.1 m)	25 ft (7.6 m)	25 ft (7.6 m)
	b) In excess of 15,000 CF (255 m³) capacity	25 ft (7.6 m)	50 ft (15.2 m)	50 ft (15.2 m)
7.	Oxygen Storage			
	a) 20,000 CF (566 m³) or less	Refer to NFPA 51, Gas Systems for Welding and Cutting		
	b) More than 20,000 CF (566 m³)	Refer to NFPA 50, Bulk Oxygen Systems at Consumer Sites		
8.	Fast-Burning Solids Such as Ordinary Lumber, Excelsior, or Paper	50 ft (15.2 m)	50 ft (15.2 m)	50 ft (15.2 m)
9.	Slow-Burning Solids Such as Heavy Timber or Coal	25 ft (7.6 m)	25 ft (7.6 m)	25 ft (7.6 m)
10.	Open Flames and Welding	25 ft (7.6 m)	25 ft (7.6 m)	25 ft (7.6 m)
11.	Air Compressor Intakes or Inlets to Ventilating or Air Conditioning Equipment	50 ft (15.2 m)	50 ft (15.2 m)	50 ft (15.2 m)
12.	Places of Public Assembly	25 ft (7.6 m)	50 ft (15.2 m)	50 ft (15.2 m)
13.	Public Sidewalks	15 ft (4.6 m)	15 ft (4.6 m)	15 ft (4.6 m)
14.	Line of Adjoining Property Which May Be Built Upon	5 ft (1.5 m)	5 ft (1.5 m)	5 ft (1.5 m)

[1] Portions of wall less than 10 ft (3 m) (measured horizontally) from any part of a system shall have a fire-resistance rating of at least 1/2 hour.

[2] Exclusive of windows and doors (see item 2).

[3] Portions of wall less than 10 ft (3 m) (measured horizontally) from any part of a system shall have a fire-resistance rating of at least 1 hour.

[4] But not less than one-half the height of adjacent wall of building or structure.

* Distances listed are from the 1999 Edition of NFPA50A.

** Distances may be reduced to 15 ft (4.5 m) for Class IIIB combustible liquids.

Location – Specific Requirements

A. Bulk gaseous hydrogen systems in excess of 15,000 standard cubic feet storage capacity must be located in a separate building or outdoors. It is preferable to locate all bulk gaseous hydrogen systems outdoors, even when the storage capacity is less than 15,000 standard cubic feet.

B. For requirements on storage of hydrogen at less than 15,000 standard cubic feet other than outdoors, see the latest edition of NFPA Code No. 50A.

C. The minimum distance in feet from a bulk gaseous hydrogen system of indicated capacity located outdoors to any specified outdoor exposure shall be in accordance with the minimum distances as given in this Safetygram.

D. If protective walls or roofs are provided, they should be constructed of noncombustible materials.

E. If the enclosing sides adjoin each other, the area should be properly vented.

F. Electrical equipment within 15 feet shall be in accordance with Article 501 or the National Electrical Code for Class 1, Division 2, Group B locations.

G. The gaseous hydrogen storage vessels and associated piping shall be electrically bonded and grounded.

H. Adequate lighting shall be provided for nighttime transfer operation.

Personnel Equipment

1. Personnel must be thoroughly familiar with the properties and safety precautions before being allowed to handle hydrogen and/or associated equipment.

2. Safety glasses, safety shoes, and leather gloves are recommended when handling cylinders.

3. In the event of emergency situations, a fire-resistant suit and gloves should be worn. Self-contained breathing apparatus (SCBA) is also recommended, but remember, atmospheres that are oxygen-deficient are within the flammable range and should not be entered.

First Aid

Persons suffering from lack of oxygen should be moved to areas with normal atmosphere. SCBA may be required to prevent asphyxiation of rescue workers. Assisted respiration and supplemental oxygen should be given if the victim is not breathing.

Fire Fighting

Hydrogen is easily ignited by heat, open flames, electrical sparks, and static electricity. It will burn with a pale blue, almost invisible flame. Most hydrogen fires will have the flame characteristic of a torch or jet and will originate at the point where the hydrogen is discharging. If a leak is suspected in any part of a system, a hydrogen flame can be detected by cautiously approaching with an outstretched broom, lifting it up and down.

The most effective way to fight a hydrogen fire is to shut off the flow of gas. If it is necessary to extinguish the flame in order to get to a place where the flow of hydrogen can be shut off, a dry powder extinguisher is recommended. However, if the fire is extinguished without stopping the flow of gas, an explosive mixture may form, creating a more serious hazard than the fire itself, should reignition occur from the hot surfaces or other sources.

The usual fire fighting practice is to prevent the fire from spreading and let it burn until the hydrogen is consumed. Dry powder fire extinguishers should be available in the area. A fire blanket should be conveniently located. An adequate water supply should be available to keep surrounding equipment cool in the event of a hydrogen fire. The local fire department should be advised of the nature of the products handled and made aware of the best known methods for combating hydrogen fires.

Pipeline fires, where shutoff is possible and with flame characteristics of a jet or torch, can be effectively controlled as follows:

1. Slowly reduce the flow of hydrogen feeding the fire. Do not completely stop the flow.

2. When the jet is small enough to be approached, put out the flame with a carbon dioxide or dry powder extinguisher.

3. Close off the supply of hydrogen completely.

4. Ventilate the area thoroughly. The probability of forming hydrogen/air mixtures within storage vessels is extremely low.

Emergency Response System

- Call: +1-800-523-9374 (Continental United States and Puerto Rico)
- Call: +1-610-481-7711 (other locations)
- 24 hours a day, 7 days a week
- For assistance involving Air Products and Chemicals, Inc. products

Product Safety Information

- For MSDS, Safetygrams, and Product Safety Information www.airproducts.com/productsafety

Technical Information Center

- Call: +1-800-752-1597 (United States)
- Call: +1-610-481-8565 (other locations)
- Fax: +1-610-481-8690
- E-mail: gasinfo@airproducts.com
- Monday–Friday, 8:00 a.m.–5:00 p.m. EST

Information Sources

- Compressed Gas Association (CGA) www.cganet.com
- European Industrial Gases Association (EIGA) www.eiga.org
- Japanese Industrial Gases Association (JIGA) www.jiga.gr.jp/english
- American Chemistry Council www.americanchemistry.com

For More Information:

Corporate Headquarters

Air Products and Chemicals, Inc.

7201 Hamilton Boulevard

Allentown, PA 18195-1501

The accuracy or completeness of all statements, technical information and recommendations contained herein is not guaranteed and no warranty of any kind is made in respect thereto. Such statements and information are given for general use only and should not be solely relied upon by the recipient when establishing appropriate procedures for his or her own operation.

appendix

Liquid Hydrogen Safety

Information created by Air Products
(www.airproducts.com/NR/rdonlyres/780E8A00-F1BF-435C-8219-
6601DA6632F0/0/safetygram9.pdf)

General

Hydrogen is colorless as a liquid. Its vapors are colorless, odorless, tasteless, and highly flammable.

Liquid hydrogen is noncorrosive. Special materials of construction are not required to prevent corrosion. However, because of its extremely cold temperature, equipment must be designed and manufactured of material that is suitable for extremely low temperature operation. Vessels and piping should be designed to the ASME Code and the ANSI Pressure Piping Code or DOT Codes for the pressure and temperatures involved.

The molecular symbol for hydrogen is H_2.

Health

Hydrogen gas is odorless and nontoxic but may produce suffocation by diluting the concentration of oxygen in air below levels necessary to support life.

Caution: The amount of hydrogen gas necessary to produce oxygen-deficient atmospheres is well within the flammable range, making fire and explosion the primary hazards associated with hydrogen and air atmospheres.

Flammability

The wide flammability range, 4% to 74% in air, and the small amount of energy required for ignition necessitate special handling to prevent the inadvertent mixing of hydrogen with air. Care should be taken to eliminate sources of ignition, such as sparks from electrical equipment, static electricity sparks, open flames, or any extremely hot objects. Hydrogen and air mixtures within the flammable range can explode and may burn with a pale blue, almost invisible flame.

Manufacture

Hydrogen is produced by the steam reforming of natural gas, the electrolysis of water and the dissociation of ammonia. Hydrogen is also a by-product of petroleum distillation and chlorine manufacture. The primary method of hydrogen generation is steam reforming of natural gas. Other feedstocks, which are less common, can include ethane, propane, butane, and light and heavy naphtha. The steam reforming process produces syngas, which is a mixture of hydrogen and carbon monoxide. Regardless of the method of production, the product stream is separated into its components. The hydrogen is dried and purified. This gaseous hydrogen is then compressed, cooled to sufficiently low cryogenic temperatures by the use of heat exchangers, reciprocating and tube expanders to form liquid hydrogen.

Uses

Liquid hydrogen is used in large volumes in the space program as a primary rocket fuel for combustion with oxygen or fluorine, and as a propellent for nuclear powered rockets and space vehicles. Although used more commonly in the gaseous state, hydrogen is stored and transported as a liquid. Hydrogen is a raw material for innumerable chemical processes ranging from the manufacturing of high-density polyethylene and polypropylene to the hydrogenation of food-grade oils. In the metallurgical industry, hydrogen is used to reduce metal oxides and prevent oxidation in heat treating certain metals and alloys. Hydrogen is also used by semiconductor manufacturers.

Containers

Liquid hydrogen is normally stored in onsite storage systems typically consisting of a tank, vaporizer and controls. Systems are selected in accordance to usage rate, pressure and regulations.

TABLE G–1. PROPERTIES

Molecular Weight	2.016
Boiling Point @ 1 atm	−423.2°F (−252.9°C)
Freezing Point @ 1 atm	−434.8°F (−259.3°C)
Critical Temperature	−400.4°F (−240.2°C)
Critical Pressure	186 psia (12.7 atm)
Density, Liquid @ B.P., 1 atm	4.42 lb./cu.ft.
Density, Gas @ 68°F (20°C), 1 atm	0.005229 lb./cu.ft.
Specific Gravity, Gas (Air = 1) @ 68°F (20°C), 1 atm	0.0696
Specific Gravity, Liquid @ B.P., [water=1 @ 68°F (20°C)]	0.0710
Specific Volume @ 68°F (20°C), 1 atm	191 cu. ft./lb.
Latent Heat of Vaporization	389 Btu/lb. mole
Flammable Limits @ 1 atm in air	4.00%–74.2% (by Volume)
Flammable Limits @ 1 atm in oxygen	3.90%–95.8% (by Volume)
Detonable Limits @ 1 atm in air	18.2%–58.9% (by Volume)
Detonable Limits @ 1 atm in oxygen	15%–90% (by Volume)
Autoignition Temperature @ 1 atm	1060°F (571°C)
Expansion Ratio, Liquid to Gas, B.P. to 68°F (20°C)	1 to 845

Tanks

Tanks are usually cylindrical in shape and placed in a horizontal position. However, some vertical cylindrical tanks and spherical tanks are in use. Tanks are mounted at fixed locations. The unit measure of capacity of tanks is the gallon. Standard tank sizes range from 1500 gallons to

25,000 gallons. Tanks are vacuum insulated. Pressure relief valves protect the tanks and are designed to ASME specifications for the pressures and temperatures involved.

Transferring Liquid

Two persons should be present when liquid hydrogen is being used or transferred or when a container is moved. This does not apply where specially trained employees of the liquid hydrogen supplier, who routinely handle liquid hydrogen, are involved.

Hydrogen is normally vaporized and used as a gas. Withdrawal of liquid from a tanker, tank, or liquid cylinder requires the use of a closed system, with proper safety relief devices, which can be evacuated and/or purged to eliminate the possibility of creating a flammable atmosphere or explosive mixture of liquid air and liquid hydrogen. Purging should be done with helium since liquid hydrogen can solidify other gases, such as nitrogen, and cause plugging and possible rupture of the transfer line or storage vessel. Liquid transfer lines must be vacuum insulated to minimize product loss through vaporization or the formation of liquid air on the lines with subsequent oxygen enrichment. All equipment must be electrically grounded and bonded before transferring liquid.

Shipment

Liquid hydrogen is transported by liquid semitrailers with a capacity of 12,000 to 17,000 gallons. The stationary tanks are filled from these tankers. Tankers are basically of the same design as the stationary tanks but in addition must meet the requirements of the Department of Transportation. Figure G–1 is a typical liquid hydrogen tanker.

FIG. G–1. TYPICAL LIQUID HYDROGEN TANKER

Safety Considerations

The hazards associated with handling liquid hydrogen are fire, explosion, asphyxiation, and exposure to extremely low temperatures. Consult the MSDS for safety information on the gases and equipment you will be using. The potential for forming and igniting flammable mixtures containing hydrogen may be higher than for other flammable gases because:

1. Hydrogen migrates quickly through small openings.

2. The minimum ignition energy for flammable mixtures containing hydrogen is extremely low.

    Burns may result from unknowingly walking into a hydrogen fire. The fire and explosion hazards can be controlled by appropriate design and operating procedures. Preventing the formation of combustible fuel-oxidant mixtures and removing or otherwise inerting potential sources of ignition (electric spark, static electricity, open flames, etc.) in areas where the hydrogen will be used is essential. Careful evacuation and purge operations should be used to prevent the formation of flammable or explosive mixtures. Adequate ventilation will help reduce the possible formation of flammable mixtures in the event of a hydrogen leak or spill and will also eliminate the potential hazard of asphyxiation. Protective clothing should be worn to prevent exposure to extremely cold liquid and cold hydrogen vapors.

3. Cold burns may occur from short contact with frosted lines, liquid air that may be dripping from cold lines or vent stacks, vaporizer fins, and vapor leaks.

Air will condense at liquid hydrogen temperatures and can become an oxygen-enriched liquid due to the vaporization of nitrogen. Oxygen-enriched air increases the combustion rate of flammable and combustible materials.

Purging

Gaseous and liquid hydrogen systems must be purged of air, oxygen, or other oxidizers prior to admitting hydrogen to the systems, and purged of hydrogen before opening the system to the atmosphere. Purging should be done to prevent the formation of flammable mixtures and can be accomplished in several ways.

Piping systems and vessels intended for gaseous hydrogen service should be inerted by suitable purging or evacuation procedures. If the piping systems are extensive or complicated, successive evacuations broken first by an inert gas and finally with hydrogen are most reliable.

Evacuating and purging of equipment in gaseous hydrogen service should include the following considerations:

1. Evacuate the equipment and break vacuum with an inert gas, such as nitrogen. Purge with an inert gas if equip ment design does not permit evacuation.

2. Repeat step 1 at least three times. If analytical equipment is available, purge system until oxygen content of residual gas is either less than or meets the process specification impurity level.

3. Hydrogen may now be introduced to the equipment.

4. Flush system with hydrogen until required purity is reached. Vent all waste hydrogen through a flue or flare stack.

Any purge method should be repeated as often as required to be certain a flammable mixture cannot be formed upon introducing hydrogen or air to the system.

Buildings

Liquid hydrogen is normally vaporized into its gaseous state and piped into buildings for usage. For storage of liquid hydrogen in a building, refer to the most recent edition of the National Fire Protection Association Pamphlet Liquefied Hydrogen Systems at Consumer Sites (NFPA No. 50B). The following items pertain to a building in which gaseous hydrogen is being used.

1. Provide adequate ventilation, particularly in roof areas where hydrogen might collect. Forced ventilation may be necessary in some applications.

2. The atmosphere in areas in which hydrogen gas may be vented and might collect should be tested with a portable or continuous flammable gas analyzer.

3. Provide an explosion venting surface or vents, taking care to vent a pressure wave to areas where people or other equipment will not become involved. Explosion vents may not be required where small quantities of hydrogen are involved.

4. Buildings should be electrically grounded.

5. Electrical equipment must conform to the existing National Electrical Codes. Electrical equip ment not conforming must be located outside the electrical area classified as hazardous. All electrical equipment must be grounded.

6. Building materials should be noncombustible.

7. Post "No Smoking" and "No Open Flames" signs.

Outdoor Storage Tank Requirements

Location—General Requirements

1. The storage containers should be located so that they are readily accessible to mobile supply equipment at ground level and to authorized personnel, and where they are not exposed to electric power lines, flammable liquid lines, flammable gas lines, or lines carrying oxidizing materials.

2. It is advisable to locate the liquefied hydrogen container on ground higher than flammable liquid storage or liquid oxygen storage. Where it is necessary to locate the liquefied hydrogen container on ground that is lower than adjacent flammable liquid storage or liquid oxygen storage, suitable protective means (such as by diking, diversion curbs, or grading) should be taken.

3. Storage sites should be fenced and posted to prevent entrance by unauthorized personnel. Sites should also be placarded as follows: "Liquefied Hydrogen Flammable Gas – No Smoking – No Open Flames."

4. Weeds or similar combustibles should not be permitted within 25 feet of any liquefied hydrogen equipment.

Location—Specific Requirements

1. The minimum distance in feet from liquefied hydrogen systems of indicated storage capacity located outdoors to any specified exposure should be in accordance with Table G–1.

2. Roadways and yard surfaces located below liquefied hydrogen piping, from which liquid air may drip, shall be constructed of noncombustible materials.

3. If protective walls are provided, they shall be constructed of noncombustible materials.

4. Electrical wiring and equipment located within three feet of a point where connections are regularly made and disconnected shall be in accordance with Article 501 of the National Electrical Code, NFPA No. 70, for Class 1, Group B, Division 1 locations.

5. Electrical wiring and equipment located beyond three feet but within 25 feet of a point where connections are regularly made and disconnected, or within 25 feet of a liquid hydrogen storage container shall be in accordance with Article 501 of the National Electrical Code, NFPA No. 70, for Class 1, Group B, Division 2 locations. This requirement does not apply to electrical equipment that is installed on mobile supply trucks or tank cars from which the storage container is filled.

6. The liquefied hydrogen container and associated piping shall be electrically bonded and grounded.

7. Adequate lighting shall be provided for nighttime transfer operations.

TABLE G–2. MINIMUM DISTANCE (FT) FROM LIQUEFIED HYDROGEN SYSTEMS TO EXPOSURES*

Type of Exposure	Total Liquefied Hydrogen Storage (Capacity in Gallons)		
	39.63 to 3500	3501 to 15,000	15,000 to 75000
1. Building/Structure			
a) Wall(s) adjacent to system constructed of noncombustible or limited combustible materials			
1) Sprinklered building/structure or unsprinklered building/structure having noncombustible contents	5[1]	5[1]	5[1]
2) Unsprinklered building/structure with combustible contents			
Adjacent wall(s) with fire-resistance rating less than 3 hours[2]	25	50	75
Adjacent wall(s) with fire resistance rating of 3 hours or greater[2]	5	5	5
b) Wall(s) adjacent to system constructed of combustible materials			
1) Sprinklered building/structure	50	50	50
2) Unsprinklered building/structure	50	75	100
2. Wall Openings			
a) Openable	75	75	75
b) Unopenable	25	50	50
3. Air Compressor Intakes, Inlets for Air-Conditioning, or Ventilating Equipment	75	75	75
4. All Classes of Flammable and Combustible Liquids (Aboveground and Vent or Fill Openings if Belowground) (See 5–1.3)[3]	75	75	75
5. Between Stationary Liquefied Hydrogen Containers	5	5	5
6. Flammable Gas Storage (other than hydrogen)	50	75	75
7. Liquid Oxygen Storage and other Oxidizers (See 5–1.3)	75	75	75
8. Combustible Solids	50	75	100
9. Open Flames and Welding	50	50	50
10. Places of Public Assembly	75	75	75
11. Public Ways, Railroads, and Property Lines	25	50	75
12. Protective Structures	5[2]	5[2]	5[2]

[1] Portion of wall less than 10 ft (3m) (measured horizontally) from any part of a system shall have a fire-resistance rating of at least 1/2 hour.
[2] Exclusive of windows and doors.
[3] Distances may be reduced to 15 ft for Class IIIB combustible liquids.
* Distances listed are from the 1999 edition of NFPA50B.
For SI Units: 1 ft = 0.305m; 1 gal = 3.785L.

Personnel Protection

1. Personnel must be thoroughly familiar with the properties and safety precautions before being allowed to handle hydrogen and/or associated equipment.

2. Full face shield, safety glasses, insulated or leather gloves, long-sleeved shirts, and pants without cuffs should be worn when working on liquid hydrogen systems. Pant legs should be worn outside of boots.

3. In the event of emergency situations, a fire-resistant suit and gloves should be worn. SCBA is also recommended, but remember that oxygen-deficient atmospheres are within the flammable range and should not be entered.

First Aid

Persons suffering from lack of oxygen should be moved to areas with normal atmosphere. SCBA may be required to prevent asphyxiation of rescue workers. Assisted respiration and supplemental oxygen should be given if the victim is not breathing.

Extensive tissue damage or burns can result from exposure to liquid hydrogen or cold hydrogen vapors. Flush affected areas with large volumes of tepid water (105–115°F, 41–46°C) to reduce freezing. Loosen any clothing which may restrict circulation. Do not apply heat. Do not rub frozen skin, as tissue damage may result. Cover affected area with a sterile protective dressing or with clean sheets if area is large, and protect area from further injury. Seek medical attention promptly.

Note to Physician

Frozen tissues should be treated promptly by immersion in a water bath at a temperature between 105–115°F (41–46°C). Avoid the use of dry heat.

Frozen tissues are painless and appear waxy with a pallid yellow color. Tissues become painful and edematous upon thawing, and the pale color turns to pink or red as circulation of blood is restored. Potent analgesics are often indicated. Tissues that have been frozen show severe, widespread cellular injury and are highly susceptible to infections and additional trauma. Therefore, rapid rewarming of tissues in the field is not recommended if transportation to a medical facility will be delayed.

If the body temperature is depressed, the patient must be warmed gradually. Shock may occur during the correction of hypothermia. Cardiac dysrhythmias may be associated with severe hypothermia.

Fire Fighting

Hydrogen is easily ignited by heat, open flames, electrical sparks, and static electricity. It will burn with a pale blue, almost invisible flame. Most hydrogen fires will have the flame characteristic of a torch or jet and will originate at the point where the hydrogen is discharging. If a leak is suspected in any part of a system, a hydrogen flame can be detected by cautiously approaching with an outstretched broom, lifting it up and down.

The most effective way to fight a hydrogen fire is to shut off the flow of gas. If it is necessary to extinguish the flame in order to get to a place where the flow of hydrogen can be shut off, a dry powder extinguisher is recommended. However, if the fire is extinguished without stopping the flow of gas, an explosive mixture may form, creating a more serious hazard than the fire itself, should reignition occur from the hot surfaces or other sources.

The usual fire fighting practice is to prevent the fire from spreading and let it burn until the hydrogen is consumed. Dry powder fire extinguishers should be available in the area. A fire blanket should be conveniently located. An adequate water supply should be available to keep surrounding equipment cool in the event of a hydrogen fire. The local fire department should be advised of the nature of the products handled and made aware of the best known methods for combating hydrogen fires.

Pipeline fires, where shutoff is possible and with flame characteristics of a jet or torch, can be effectively controlled as follows:

1) Slowly reduce the flow of hydrogen feeding the fire. Do not completely stop the flow.

2) When the jet is small enough to be approached, put out the flame with a carbon dioxide or dry powder extinguisher.

3) Close off the supply of hydrogen completely.

4) Ventilate the area thoroughly. The probability of forming hydrogen/air mixtures within storage vessels is extremely low.

Emergency Response System

- Call: +1-800-523-9374 (Continental United States and Puerto Rico)
- Call: +1-610-481-7711 (other locations)
- 24 hours a day, 7 days a week
- For assistance involving Air Products and Chemicals, Inc. products

Product Safety Information

- For MSDS, Safetygrams, and Product Safety Information www.airproducts.com/productsafety

Technical Information Center

- Call: +1-800-752-1597 (United States)
- Call: +1-610-481-8565 (other locations)
- Fax: +1-610-481-8690
- E-mail: gasinfo@airproducts.com
- Monday–Friday, 8:00 a.m.–5:00 p.m. EST

Information Sources

- Compressed Gas Association (CGA) www.cganet.com
- European Industrial Gases Association (EIGA) www.eiga.org
- Japanese Industrial Gases Association (JIGA) www.jiga.gr.jp/english
- American Chemistry Council www.americanchemistry.com

For More Information:

Corporate Headquarters

Air Products and Chemicals, Inc.

7201 Hamilton Boulevard

Allentown, PA 18195-1501

The accuracy or completeness of all statements, technical information and recommendations contained herein is not guaranteed and no warranty of any kind is made in respect thereto. Such statements and information are given for general use only and should not be solely relied upon by the recipient when establishing appropriate procedures for his or her own operation.

Glossary and Conversion Factors

Glossary

absorption
: A process in which a gas or liquid is drawn into another substance to form a compound. An example is the absorption of gaseous hydrogen by a metal, forming a hydride.

a-c
: Alternating-current. Electrical current in which the electrons flow in both directions, from positive to negative and from negative to positive, in the same conductor.

acidic
: A solution with a pH level below 7.0. Acids are the opposite of basic or alkaline solutions.

adsorption
: The adhesion of molecules to the surface area of solids or liquids. For example, gaseous hydrogen is adsorbed onto porous ceramic materials.

AFC
: Alkaline fuel cell. A fuel cell technology that uses potassium hydroxide (KOH) or other alkaline solution as an electrolyte.

alkaline	A solution with a pH higher than 7.0. Alkaline solutions, also called basic solutions, are the opposite of acids.
alternative fuel	A popular term for unconventional fuels. Typically, alternative fuels are derived from natural gas, biomass, or other non-polluting or less-polluting sources such as hydroelectric, wind, or solar power.
alternator	A device that produces alternating-current electricity. Also called a synchronous generator or a-c generator.
anion	A negatively charged ion.
ANL	Argonne National Laboratory (U.S.)
anode	One of two electrodes in an electrochemical cell. In a fuel cell, the anode is electrically negative. In an electrolysis system, the anode is electrically positive.
ANSI	American National Standards Institute
ARB	Air Resources Board (California)
ASME	American Society of Mechanical Engineers
avoided cost	The price utilities pay for electricity produced by cogenerating micropower plants. The utility's avoided cost equals whatever it would have cost the utility to generate the power.
bar	Metric unit of pressure, or force per unit of area, equal to 100,000 Pascals.
baseload	The primary source of round-the-clock electrical power supply; the minimum required at a steady rate over a specified period of time.

basic	A solution with a pH higher than 7.0. Basic solutions, also called alkaline solutions, are the opposite of acids.
bcm	Billion cubic meters. Metric unit that measures volume.
biomass	Biomass is organic material of various types including inedible plants, undesirable species that create a nuisance, crops grown specifically as biomass crops, leftovers from agricultural crops, organic industrial waste such as wood chips from sawmills, animal waste from farming operations, solid waste collected in urban areas, and biogases generated from landfills and sewage treatment plants.
black hydrogen	Hydrogen produced from coal or nuclear power rather than from clean, renewable power sources.
borate	A class of boron-oxygen compounds.
Borax	A type of borate and a common ingredient in laundry detergents.
boron	A relatively light element (atomic weight of 10.8) represented by the symbol "B" on the periodic table.
Btu	British thermal unit. Unit that measures heat, equal to the amount required to raise the temperature of one pound of water by one degree Fahrenheit.
buyback rate	The price at which utilities purchase electricity from third parties such as operators of micropower plants.
by-product	A substance produced from a process that is not the main product. For example, when a fuel cell uses hydrogen and oxygen to produce electricity, the by-product is water.

C_3H_8 Propane. A component of natural gas and the main component of liquefied petroleum gas. Propane is usually extracted from natural gas and sold separately.

CaFPC   California Fuel Cell Partnership

CARB   California Air Resources Board

carbon   A relatively light element (atomic weight of 12) represented by the symbol "C" on the periodic table. Carbon is a constituent in all hydrocarbons and other organic compounds.

carbon-neutral   A process that could result in no net emissions of carbon, such as biomass conversion.

catalyst   A chemical that increases the rate of a reaction without being consumed. The catalyst lowers the activation energy required, allowing the reaction to proceed more quickly or at a lower temperature. In a fuel cell, catalysts facilitate the reaction of oxygen and hydrogen.

catalyst poisoning   A process in which impurities bind to a catalyst, reducing its ability to facilitate the desired chemical reaction.

cathode   One of two electrodes in an electrochemical cell. In a fuel cell, the cathode is electrically positive. In an electrolysis system, the cathode is electrically negative.

cation   A positively charged ion.

CBO   Congressional Budget Office (U.S.)

cell	An individual electrochemical unit consisting of two electrodes and an electrolyte. Examples include fuel cells and electrolytic cells.
CFCs	Chlorofluorocarbons. Chemical compounds containing carbon, chlorine, and fluorine which were found to deplete the earth's protective ozone layer.
CH2IP	Compressed Hydrogen Infrastructure Program (Canada)
CH_3CH_2OH	Ethanol, or ethyl alcohol. A clear, colorless liquid that is the same alcohol found in beer, wine, and liquor. Ethanol is produced by fermenting a sugar solution with yeast.
CH_3OH	Methanol, or methyl alcohol. Methanol, the simplest of all alcohols, is a flammable, poisonous liquid used as a fuel, fuel additive, chemical feedstock, and potential energy carrier.
CH_4	Methane. A colorless, odorless, flammable gas that is the main constituent of natural gas.
CHP	Combined heat and power. A system that generates electricity and recovers heat for useful purposes. Also called cogeneration. Fuel cells can be applied in CHP systems.
climate change	The warming of the earth's climate over millions of years, which is thought to be caused by carbon dioxide and other greenhouse gases in the atmosphere. Also called global warming.
CO	Carbon monoxide. A colorless, odorless gas that is also poisonous. Carbon monoxide results from incomplete combustion of carbon with oxygen.

CO_2 — Carbon dioxide. A colorless, odorless gas produced by combustion of hydrocarbon fuels, the respiration of living organisms, and other processes. Carbon dioxide is a greenhouse gas and a major contributor to global warming.

cogeneration — A system that generates electricity and recovers heat for useful purposes. Also called combined heat and power, or CHP. Fuel cells can be applied in cogeneration systems.

combustion — The process of burning, also called oxidation. Combustion occurs when a fuel or other substance combines with oxygen or air in the proper proportions and in the presence of a spark or other heat source.

composite — A substance created by combining different materials to obtain certain properties or characteristics. Each constituent retains its physical identity.

cryogenic — Very low temperatures, often those at which gases change into liquid form.

CSIRO — Commonwealth Scientific and Industrial Research Organization (Australia)

CSLF — Carbon Sequestration Leadership Forum

CTFCA — Canadian Transportation Fuel Cell Alliance

CUTE — Clean Urban Transport for Europe. A demonstration project that operates 30 fuel cell buses, three in each of 10 cities.

d-c — Direct-current. Electrical current in which the electrons flow in only one directions, from positive to negative or from negative to positive.

demand charge	Charges for the use of electricity based on the maximum power consumed during a specified period of time.
density	Mass per unit of volume. Density varies with temperature and pressure.
DG	Distributed generation. Power that is generated at or near the customer's site, usually in relatively small amounts. Also called micropower.
DOE	Department of Energy (U.S.)
DOT	Department of Transportation (U.S.)
$/kg-d	Cost in dollars per kilogram of hydrogen produced per day.
EC	European Commission
Ectos	Ecological City Transport System (Iceland)
EE	Eastern Europe
EEA	European Environment Agency
EERE	Energy Efficiency and Renewable Energy (office within the U.S. Department of Energy)
EIA	Energy Information Administration (U.S.)
electrode	One of two electronically conducting parts of an electrochemical cell, such as a fuel cell or electrolyzer. An electrode can be a rod, a sheet, or a more complex structure.
electrolysis	An electrochemical process in which an electrical current passes through an electrolyte, causing a chemical reaction. For example, electrolysis splits the water molecule into hydrogen and oxygen.

electrolyte	A substance that conducts charged ions from one electrode to the other inside a fuel cell or electrolyzer.
electrolyzer	An electrochemical device that produces hydrogen and oxygen from water.
electron	An atomic particle that has a negative electrical charge. Electricity is the flow of electrons.
endothermic	A process that requires heat to take place.
energy carrier	A liquid or gas in which energy can be stored in a chemical form and transported. Also a non-material carrier such as electricity or solar radiation.
energy density	Amount of energy in a given volume.
EPA	Environmental Protection Agency (U.S.)
EPRI	Electric Power Research Institute (U.S.)
ethanol	See CH_3CH_2OH.
EU	European Union
EUR	Estimated ultimate recovery
exothermic	A process that gives off heat when it takes place.
FCV	fuel cell vehicle
flammability limits	The range of flammability in which a gas and air are in the right proportions to burn when ignited. Below the lower limit, there is not enough fuel to burn. Above the higher limit, there is not enough oxygen to support combustion.
FSU	Former Soviet Union
fuel cell stack	Individual fuel cells connected in series. Increasing the stack's number of cells boosts its voltage.

fuel cell	An electrochemical cell that converts the chemical energy contained in a fuel (such as hydrogen) directly into electricity through a reaction with oxygen. The electrochemical reaction is flameless.
fuel processor	Device used to extract hydrogen from hydrocarbon fuels, usually by reforming natural gas. Some fuel processors can accept other fuels such as propane, ethanol, methanol, or gasoline.
gge	Gallon of gasoline equivalent. The quantity of fuel such as hydrogen which contains the same amount of energy as a gallon of gasoline.
GHG	Greenhouse gas. These gases, which include carbon dioxide and methane, are thought to cause global warming.
global warming	A gradual process in which greenhouse gases in the earth's atmosphere allow the sun's rays in, but prevent heat from being radiated back into space.
grid	The network of equipment that transmits and distributes electricity.
GW	Gigawatt. Metric unit that measures power, equal to one billion Watts or 1,000 megawatts.
GW_e	Gigawatt–electrical
GW_{th}	Gigawatt–thermal
H_2	Hydrogen. The lightest element (atomic weight of 1). Hydrogen is a colorless, odorless, flammable gas that liquefies at cryogenic temperatures ($-253°C$, or $-423°F$). Hydrogen forms water when burned or when combined with oxygen in a fuel cell.

h2EA | h2 Early Adopters. A Canadian government-industry program to demonstrate hydrogen technologies that further the transition toward a hydrogen economy in Canada.

heat exchanger | A mechanical device such as a radiator that is designed to move heat from one part of a system to another or from one mass to another mass.

heating value | The amount of energy produced by combustion of a fuel. See HHV and LHV.

HHV | Higher heating value. The amount of energy produced by combustion of a fuel, including the heat given off when water vapor in the combustion products condenses.

HOD | Hydrogen on Demand™, a type of hydrogen generation system.

hp | Horsepower. Unit that measures the rate of doing work, equal to 746 Watts.

hydrocarbon | An organic fuel containing hydrogen and carbon alone or in combination with other elements. Hydrocarbon fuels include coal, oil, gasoline, methane, and propane.

hydrocracking | A method of breaking down complex hydrocarbons into their various components, using hydrogen and a catalyst.

hydrogen economy | An economic system that uses hydrogen as the primary or secondary energy carrier.

hydrogenation | The process of chemically combining with hydrogen. For example, edible oils are hydrogenated to prolong their shelf life.

IEEE	Institute of Electrical & Electronics Engineers
IGCC	Integrated gasification combined cycle. A system that integrates coal gasification with two turbines operating in a combined cycle. The synthesis gas produced from coal is burned as fuel in a combustion turbine-generator. The turbine's exhaust heat is used in a boiler to create steam that powers a second turbine-generator.
inorganic	A compound that does not contain carbon.
ion	Atom or molecule that carries a positive or negative charge because of the loss or gain of electrons.
IPCC	Intergovernmental Panel on Climate Change (United Nations)
IPHE	International Partnership for the Hydrogen Economy
ISO	International Organization for Standardization
J	Joule. Metric unit that measures the amount of work done or energy expended by a force of 1 Newton acting through a distance of 1 meter.
JHFC	Japan Hydrogen and Fuel Cell demonstration program
kg	Kilogram. Metric unit that measures mass or weight, equal to about 2.2 pounds.
kg/d	Kilograms per day
kg/yr	Kilograms per year
kPa	Kilopascal. Metric unit that measures pressure, equal to 1,000 Pascals.

kW	Kilowatt. Metric unit that measures power, equal to 1,000 Watts or about 1.34 horsepower.
kWh	Kilowatt-hour. Metric unit that measures electrical energy.
LANL	Los Alamos National Laboratory (U.S.)
LH_2	Liquid or liquefied hydrogen
LHV	Lower heating value. The amount of energy produced by combustion of a fuel, not including the heat given off when water vapor in the combustion products condenses.
LLNL	Lawrence Livermore National Laboratory (U.S.)
LNG	liquefied natural gas
LPG	Liquefied petroleum gas. A mixture of hydrocarbon gases such as propane and butane that have been liquefied for storage and use.
m^3	Cubic meters. Metric unit that measures volume.
MCFC	Molten carbonate fuel cell. A type of high-temperature fuel cell that uses molten carbonate salt mixture as an electrolyte, suspended in a porous, chemically inert ceramic lithium-aluminum oxide matrix.
membrane	A layer of thin film that acts as an electrolyte in fuel cells and electrolyzers.

metal hydride   A compound created when a metal or alloy absorbs gaseous hydrogen. In metal hydride storage technology, hydrogen fills the spaces in the metal's crystalline lattice structure, but occupies no additional space. When the metal hydride is heated, hydrogen is released.

methane   See CH_4.

methanol   See CH_3OH.

METI   Ministry of Economy, Trade and Industry (Japan)

microgrid   A group of homes or larger buildings that are interconnected and serviced by multiple micropower units.

micropower   Any small electricity generation unit (less than 20 megawatts), usually located at or near the power customer's site. Hydrogen fuel cells and other small power generators can be used in micropower plants. Also called distributed generation.

microturbine   A very small turbine based on automotive turbocharger technology with output ranging from 25 to 300 kilowatts.

mJ   Megajoule. Metric unit that measures energy, equal to one million Joules.

MPa   Megapascal. Metric unit that measures pressure, equal to one million Pascals.

mpg   miles per gallon

MW   Megawatt. Metric unit that measures power, equal to one million Watts.

MW_e	Megawatt-electric. Metric unit that measures electrical power.
MW_{th}	Megawatt-thermal. Metric unit that measures thermal power.
NAE	National Academy of Engineers (U.S.)
NAS	National Academy of Sciences (U.S.)
NASA	National Aeronautics and Space Administration (U.S.)
natural gas	A naturally occurring mixture of hydrocarbon gases, primarily methane (CH_4).
neutron	An atomic particle that has no electrical charge.
Newton	Metric unit that measures the amount of force that produces an acceleration of 1 meter per second when exerted on a mass of 1 kilogram.
NGLs	Natural gas liquids. Hydrocarbons heavier than methane that are found in some natural gas, including condensate, butane, propane, and ethane.
nm^3	Normal cubic meters. Metric unit that measures volume.
NO_x	Nitrogen oxides. Chemical compounds of nitrogen and oxygen that are created at high temperatures during the combustion of fuels. Nitrogen oxides contribute to urban smog and can cause acid rain.
NRC	National Research Council (U.S.)
NREL	National Renewable Energy Laboratory (U.S.)
OPEC	Organization of Petroleum Exporting Countries
organic	A compound containing carbon

OTEC   Ocean thermal energy conversion. A technology that exploits the temperature difference between the warm surface of the sea and its colder depths.

oxidation   The process of combining with oxygen, accompanied by the loss of one or more electrons from the atom or molecule being oxidized. Combustion, or burning, is a type of oxidation, but oxidation can also take place electrochemically without a flame as in a fuel cell. The rusting of metals is a slow oxidation process.

Pa   Pascal. Metric unit that measures pressure, or force per unit area. One Pascal is the pressure generated by a force of 1 Newton acting on an area of 1 square meter. It is a small unit and is more often used as a kilopascal (kPa) or megapascal (MPa).

PAFC   Phosphoric acid fuel cell. A type of fuel cell that uses concentrated liquid phosphoric acid (H_3PO_4) as an electrolyte—the acid is contained in a Teflon-bonded silicon carbide matrix—and porous carbon electrodes containing a platinum catalyst.

partial oxidation   A type of reforming reaction in which some of the feedstock is burned as fuel to generate the heat required for subsequent reactions, but combustion is quenched before it is complete.

particulates   Sooty particles produced from burning coal, diesel fuel, and other heavy hydrocarbons.

peak demand period   Daily and seasonal times of maximum electricity demand.

peak power   Electricity supplied during peak demand periods.

peakshaving	Reducing consumption of electricity during peak demand periods.
PEM	Polymer electrolyte membrane or proton exchange membrane. A thin plastic film that serves as the electrolyte in a fuel cell or electrolyzer.
PEM electrolyzer	A type of electrolyzer that uses a solid polymer (thin plastic film) as an electrolyte.
PEMFC	PEM fuel cell. A type of low-temperature fuel cell that uses a solid polymer (thin plastic film) as an electrolyte and porous carbon electrodes containing a platinum catalyst or other noble metal.
PNNL	Pacific Northwest National Laboratory (U.S.)
primary energy	Sources of energy found in nature such as coal, oil, and natural gas.
propane	See C_3H_8
proton	An atomic particle that has a positive electrical charge.
PSA	Pressure swing adsorption. A process for removing carbon dioxide from a gaseous stream.
psi	Pounds per square inch. Unit that measures pressure, or force per unit area.
psig	Pounds per square inch gauge
PV	Photovoltaic. Systems of solar cells that convert the sun's light energy directly into an electric current. In photovoltaic cells, sunlight is absorbed by semiconducting materials such as crystalline silicon, allowing electrons from their atoms to flow through an external circuit and generate electricity.

real-time pricing   Electricity rate structure that charges premium prices for power during peak demand periods. Also called time-of-day rates.

reformer   A device in which hydrocarbon fuels or feedstocks are reacted with steam in the presence of a catalyst to produce a hydrogen-carbon monoxide synthesis gas from which hydrogen is separated and purified.

reforming   A chemical process that uses steam and a catalyst to convert hydrocarbon fuels or feedstocks into a hydrogen-carbon monoxide synthesis gas from which hydrogen is separated and purified.

renewable energy   Energy derived from resources that are virtually unlimited because they are renewed by nature within a short time. Examples include solar energy, wind power, and hydroelectric power.

scf   Standard cubic feet. Unit that measures volume.

shift reaction   One of the steps in reforming natural gas or other hydrocarbons. The shift reaction changes the carbon monoxide to carbon dioxide and produces more hydrogen.

SMR   steam methane reforming

SNL   Sandia National Laboratories

sodium borohydride   An advanced chemical storage material that combines hydrogen with sodium borate (essentially, the familiar laundry detergent Borax). The chemical creates a nontoxic, nonflammable solution with water and produces hydrogen when exposed to a catalyst. When the catalyst and solution are separated, the system stops generating hydrogen.

SOFC   Solid oxide fuel cell. A type of fuel cell that uses a hard, non-porous ceramic compound as the electrolyte. They operate at very high temperatures—around 1,000°C (1,800°F), which means they do not require an expensive precious-metal catalyst. High-temperature operation also allows solid oxide cells to reform fuels internally.

SO_x Sulfur oxides. Compounds containing sulfur and oxygen. Sulfur oxides, created by combustion of coal and other sulfurous fuels, can cause acid rain.

syngas   A synthesis gas containing hydrogen and carbon monoxide that is produced during the reforming of natural gas or other hydrocarbons.

tcf   Trillion cubic feet. Unit that measures volume.

time-of-day rates   Electricity pricing structure that charges premium rates for power during peak demand periods. Also called real-time pricing.

ton   U.S. unit of weight equal to 2,000 pounds.

tonne   Metric unit that measures weight, equal to 1,000 kilograms or about 2,200 pounds.

tpy   Tons or tonnes per year

turbocharger   A compressor used to increase the pressure and density of gases entering an engine. The compressor is driven by a turbine that extracts energy from the engine's exhaust gas.

U.N.   United Nations

UPS   Uninterruptible power supply

UV-IR   Ultraviolet-infrared detector

VOCs   Volatile organic compounds. Compounds that have a high vapor pressure and low water solubility and which contribute to urban smog. Many VOCs are artificial chemicals used and produced in the manufacture of paints, pharmaceuticals, and refrigerants. VOCs typically are industrial solvents such as trichloroethylene and fuel additives such as methyl tertiary butyl ether (MTBE).

W   Watt. Metric unit that measures power, or the rate of doing work. One Watt equals 1 Joule per second.

We-Net   World Energy Network (Japan)

wheeling   Third-party transportation of power on utility transmission lines.

ZEV   Zero-emission vehicle. A vehicle that has no tailpipe emissions, no evaporative emissions, no emissions from gasoline refining or sales, and no onboard emission-control systems that can deteriorate over time.

CONVERSION FACTORS

Distance		
1 mile	=	1.609347 km
1 km	=	0.62137 mile
Volume		
1 scf	=	0.028317 nm³
1 nm³	=	35.3145 scf
1 gallon	=	3.785 liters
1 liter	=	0.264 gallon
1 m³	=	1,000 liters
Vehicle fuel use		
1 mpg	=	0.4251 km/liter
1 km/liter	=	2.352 mpg
Mass		
1 kg	=	2.2046 lb
1 lb	=	0.4536 kg
1 ton	=	0.978 tonne
1 metric ton	=	1.023 tons
	=	2,204.6 lb
	=	1,000 kg
Pressure		
1 psi	=	0.06895 bar
	=	6.895 kPa
1 bar	=	14.5 psi
	=	100 kPa
	=	0.1 MPa
1 MPa	=	145 psi
Temperature		
delta °F	=	1.8°C
delta °C	=	0.555°F
°F	=	1.8°C + 32
°C	=	(°F − 32)0.555
Energy/Work		
hp	=	746 W
	=	0.746 kW
kW	=	1.34 hp
	=	1,000 W
MW	=	1,000,000 W
Btu	=	1,055 J
Btu	=	1.055 kJ
kJ	=	0.948 Btu
kWh	=	3,600,000 J
	=	3,600 kJ
	=	3.6 MJ
Energy content		
1 gallon gasoline	=	1 kg H_2
	=	380 scf H_2
	=	10 nm³ H_2
1 kg H_2	=	1 gallon gasoline
100 scf H_2	=	9.30 kWh HHV
	=	7.87 kWh LHV
	=	31,740 Btu HHV
	=	26,840 Btu LHV

Bibliography

Books and Other Publications

Al-Jubeir, Adel, in Wessel, David, "Capital: Saudi Arabia Fears $40-a-Barrel Oil, Too," *The Wall Street Journal*, May 27, 2004.

Aldefer, R. Brent; Eldridge, M. Monika; and Starrs, Thomas J., *Making Connections: Case Studies of Interconnection Barriers and Their Effect on Distributed Power Projects.* NREL/SR-200-28053. Golden, CO: National Renewable Energy Laboratory, May 2000.

American Wind Energy Association, "Wind Energy Comes of Age," www.awea.org, March 25, 2004.

Anderson, Patrick L., and Geckil, Ilhan K., "Northeast Blackout Likely to Reduce US Earnings by $6.4 Billion," AEG Working Paper 2003-2, August 19, 2003.

Asia Pulse Pte Ltd., "Japan's Fuji Electric Develops Fuel Cell That Runs 7.5 Years," *Fuel Cell Today*, June 22, 2004, www.fuelcelltoday.com.

Baard, Mark, "Hydrogen Is No Gas, Yet," *Wired*, June 23, 2003, www.wired.com.

Baard, Mark, "Hydrogen's Dirty Details," *Village Voice*, January 6, 2004, www.villagevoice.com.

Baard, Mark, "Raising a Stink, With Hydrogen," *Wired*, July 4, 2003, www.wired.com.

Ball, Jeffrey, "Industries Seek Fuel-Cell Funding," *The Wall Street Journal*, November 18, 2003.

Bennett, Jeff, "An Experiment: GM, Shell plan hydrogen pump in D.C. area for 6 minivans' use," *Detroit Free Press*, March 5, 2003.

"The Big Mac Index: Food for Thought," *The Economist*, Volume 371, Number 8377, May 29-June 4, 2004, pages 71–72.

"Blackout Cost New York $36m an Hour," Guardian Newspapers Ltd., August 19, 2003.

Blankinship, Steve, "When Will the Hydrogen Future Arrive?", *Power Engineering*, January 2003, http://pe.pennwellnet.com.

Blankinship, Steve, "Fuel Cell Component Manufacturer Discusses Status and Future of Fuel Cell Industry", *Power Engineering*, April 2003, http://pe.pennwellnet.com.

Bossel, Ulf, "The Physics of the Hydrogen Economy," *European Fuel Cell News*, Volume 10, Number 2, July 2003.

Braginton-Smith, Brian, "Ocean Based Renewable Energy," www.energypulse.net.

Brooks, Bob, "GM, Dow to Begin Fuel-Cell Power-Generation Project," July 8, 2003, WardsAuto.com.

Business Communications Co., "10–15% of Laptops Will Run on Micro Fuel Cells in 2012, Forecasts ABI Research," May 12, 2004.

Business Communications Co., "The Power Quality Equipment and Services Market: A Growing 21st Century Business," November 2001.

Chambers, Ann. *Distributed Generation: A Nontechnical Guide.* Tulsa, OK: PennWell, 2001.

Cheang, Michael, "Powered by the sun and hydrogen," April 20, 2004, www.thestar.com.my.

"China Report," *Hydrogen & Fuel Cell Letter*, June 2004, www.hfcletter.com.

Congressional Budget Office, "Prospects for Distributed Electricity Generation," September 2003, www.cbo.gov.

DE Solutions, Inc., *Clean Distributed Generation Performance and Cost Analysis.* Report for Oak Ridge National Laboratory and U.S. Department of Energy, April 2004.

"Depleting Oil Reserves Push European Renewables," RE *Weekly News*, July 9, 2004, www.solaraccess.com.

Dickson, Edward M., Ryan, John W., and Smulyan, Marilyn H., *The Hydrogen Energy Economy: A Realistic Appraisal of Prospects and Impacts.* Praeger Publishers, Inc.: New York, 1977. ISBN 0-275-24290-0. Chapters 1, 2

Duncan, Richard C., "Three world oil forecasts predict peak production," *Oil & Gas Journal*, May 26, 2003, pp. 18–21.

Dunn, Seth, *Hydrogen Futures: Toward a Sustainable Energy System* (*Worldwatch Paper* #157). Worldwatch Institute, August 2001. ISBN 1-878071-59-9.

Dunn, Seth, in "Hydrogen: The New Power Source?", www.satyamag.com, December 2001.

"The Economic Cost of the Blackout," ICF Consulting, August 14, 2003.

Edison Electric Institute, "Industry Statistics," May 2003, www.eei.org.

Electric Power Research Institute, "Distributed Energy Resources Market is Starting to Rebound, Says Primen: New Study Identifies Key Factors Influencing Purchase Decisions," January 26, 2004, www.epri.com.

Electricity Consumers Resource Council (ELCON), "The Economic Impacts of the August 2003 Blackout," February 9, 2004.

Eliasson, Baldur, and Bossel, Ulf, "The Future of the Hydrogen Economy: Bright or Bleak?" Paper presented at Fuel Cell World, Lucerne, July 5, 2000.

"Energy: These Fuelish Things," *The Economist*, August 16, 2003, pages 56-57.

European Commission, "Renewable energy: Commission calls for a stronger commitment of Member States to achieve the 2010 targets," May 26, 2004, http://europa.eu.int.

European Environment Agency, "Domestic measures taken or planned so far are insufficient to meet EU climate emissions targets, projections show," December 2, 2003, http://org.eea.eu.int.

European Union, "EU announces plan for Hydrogen Economy," October 24, 2002.

Evers, Arno A., "Fueling our Future: Setting the Stage for the Coming Hydrogen Economy." Paper presented at the 15th World Hydrogen Energy Conference, Yokohama, June 27–July 2, 2004.

Evers, Arno A., "Why Should I Buy a Fuel Cell?", February 7, 2003, www.energypulse.net.

"Fuel Cells: A new kind of gas station," *The Economist*, December 6, 2003, page 59.

Geiger, Stefan, "Fuel Cell Market Survey: Automotive Hydrogen Infrastructure," *Fuel Cell Today*, May 5, 2004, www.fuelcelltoday.com.

Geiger, Stefan, "First CUTE fuel cell bus arrives in Germany," *Fuel Cell Today*, September 16, 2003, www.fuelcelltoday.com.

Glover, Mark, "Hydrogen-powered Fords set for test-drive in capital," *Sacramento Bee*, April 28, 2004, www.sacbee.com.

"GM Well-to-Wheel Analysis of Energy Use and Greenhouse Gas Emissions of Advanced Fuel/Vehicle Systems—A European Study," September 27, 2002, www.lbst.de/gmwtw.

Goswami, Darshan, "Hydrogen Economy—A Revolutionary Vision for the Future of Energy," February 4, 2003, www.energypulse.net.

Goswami, Darshan, "Fuel Cells to Revolutionize Electric Power Generation," April 1, 2003, www.energypulse.net.

Guynn, Mark D., Freeh, Joshua E., and Olson, Erik D., "Evaluation of a Hydrogen Fuel Cell Powered Blended-Wing-Body Aircraft Concept for Reduced Noise and Emissions," Report No. NASA/TM-2004-212989, February 2004.

Hoffmann, Peter, *Tomorrow's Energy: Hydrogen, Fuel Cells, and the Prospects for a Cleaner Planet*. The MIT Press: Cambridge, MA, 2001. ISBN 0-262-08295-0.

Hoffmann, Peter, "The Hydrogen Power Rush," 2002, www.worldandi.com.

"Hydrogen Economy Breakthrough on the Horizon," February 13, 2004, www.solaraccess.com.

"Hydrogen Market to Nearly Double To US$1.5 Billion By 2008," *Fuel Cell Today*, www.fuelcelltoday.com, 19 January 2004.

"IEEE Distributed Energy Interconnection Standard Published," EPRI *Journal*, August 20, 2003, www.epri.com.

Illinois 2H$_2$ Partnership, "The Hydrogen Highway: Illinois' Path to a Sustainable Economy and Environment," March 2004, www.ilcoalition.org.

"Illinois Coalition Wins Grant to Develop Solar and Wind Powered Hydrogen Station in Rockford," *Fuel Cell Today*, May 20, 2004, www.fuelcelltoday.com.

"Industry members optimistic about next 50 years for nuclear energy," *Power Engineering*, May 22, 2003.

Intergovernmental Panel on Climate Change (United Nations), IPCC *Third Assessment Report—Climate Change* 2001, 2001, www.ipcc.ch

International Energy Agency, "Distributed-Generation Issues in Japan, the US, the Netherlands, and the UK," www.iea.org.

Ivy, Johanna, "Summary of Electrolytic Hydrogen Production: Milestone Report for the Department of Energy's Hydrogen, Fuel Cells, and Infrastructure Technologies Program," National Renewable Energy Laboratory Technical Report NREL/MP-560-35948, April 2004.

Jenkins, Holman W. Jr., "Nothing to Say? Talk About 'Energy Independence'," *The Wall Street Journal*, June 2, 2004, page A15.

Labs, Wayne, "Fuel Cells Go Back to the Future—Again," *Electronic Design*, June 14, 2004, pages 83-85.

Lloyd, Andrew C., "The Power Plant in Your Basement," *Scientific American*, July 1999.

Lord Browne, "Climate Change," Institutional Investors Group, Bishopsgate, London, November 26, 2003, www.bp.com.

Maugeri, Leonardo, "Time to debunk mythical links between oil and politics," *Oil & Gas Journal*, December 15, 2003, pages 18-28.

McNeill, John R., *Something New Under the Sun: An Environmental History of the Twentieth-Century World*. New York: Norton, 2000.

Mintz, Marianne, Folga, Stephen, Molburg, John, and Gillette, Jerry, "Cost of Some Hydrogen Fuel Infrastructure Options," Argonne National Laboratory, January 16, 2002, www.anl.gov.

Mogford, John, "Growing Markets for Cleaner Energy: The Role of Hydrogen within an Energy Major," Hyforum Conference, Beijing, May 25, 2004, www.bp.com.

Moritis, Guntis, "CO_2 sequestration adds new dimension to oil, gas production," *Oil & Gas Journal*, March 3, 2003, pages 39-44.

National Research Council, *The Hydrogen Economy: Opportunities, Costs, Barriers, and R&D Needs*. The National Academies Press, 2004 (394 pages). ISBN 0-309-09163-2, www.nap.edu.

National Academy of Sciences, "Climate Change Science: An Analysis of Some Key Questions," 2001, www.nap.edu.

National Research Council, "Reconciling Observations of Global Temperature Change," January 2000.

Parks, Bill, U.S. Department of Energy, "Transforming the Grid to Revolutionize Electric Power in North America." Paper presented at the Edison Electric Institute Fall 2003 Transmission, Distribution and Metering Conference, October 13, 2003.

Parrish, Alton, "Hydrogen Market to Nearly Double to US$1.5 Billion By 2008," *Fuel Cell Technology News*, January 19, 2004.

Peavey, Michael A., *Fuel from Water: Energy Independence with Hydrogen*, 11th edition. Merit, Inc.: Louisville, KY, 2003. ISBN 0-945516-04-5.

"Plug Power and Honda demonstrate Home Energy Station," *Fuel Cell Today*, October 2, 2003, www.fuelcelltoday.com.

Pope, Hugh, and Cummins, Chip, "Saudis Suffer Fresh Terrorist Attack," *The Wall Street Journal*, June 1, 2004, page A3.

Powers, Laurie, "Flexibly Fueled Storage Tank Brings Hydrogen-Powered Cars Closer to Reality," Lawrence Livermore National Laboratory, June 2003, www.llnl.gov.

Rifkin, Jeremy, *The Hydrogen Economy: The Creation of the Worldwide Energy Web and the Redistribution of Power on Earth*. Penguin Group (USA) Inc.: New York, 2002. ISBN-1-58542-193-6.

Samotyj, Marek, "The Cost of Power Disturbances to Industrial & Digital Economy Companies." Report by Primen for Electric Power Research Institute, Consortium for an Electric Infrastructure to Support a Digital Society (CEIDS), June 2001.

Samsam Bakhtiari, A. M., "Middle East oil production to peak within next decade," *Oil & Gas Journal*, July 7, 2003, pages 20-28.

"Saudi Arabia and Oil: What If?," *The Economist*, Volume 371, Number 8377, May 29–June 4 2004, pages 69–70.

Schroeder, Chip, "Hydrogen from Electrolysis," RE *Weekly News*, June 21, 2004, www.solaraccess.com.

Schubert, Kevin, "Fuel Cell DG: A Stationary Solution to Mobilizing the Hydrogen Economy," www.energypulse.net, November 17, 2003.

Silberman, Steve, "Girding Up For the Power Grid," *Wired*, June 14, 2001, www.wired.com.

Silberman, Steve, "The Energy Web," *Wired*, Issue 9.07, July 2001, www.wired.com.

Siemens Westinghouse, www.siemenswestinghouse.com.

Silverstein, Ken, "Fuel Cells Get Big Lift," Issue Alert, May 9, 2003, www.utilipoint.com.

Silverstein, Ken, "Hybrid Cars on Fast Track," Issue Alert, May 25, 2004, www.utilipoint.com.

Smalley, Richard, "The Future of Fuel: Advances in Hydrogen Fuel Cell Technology," PBS Online NewsHour, October 20, 2003, www.pbs.org.

Sperling, Daniel, and Ogden, Joan, "The Hope for Hydrogen," *Issues in Science and Technology*, Spring 2004, pages 82-86.

Travers, Bridget, Ed., *World of Invention*. Gale Research Inc.: Detroit, 1994. ISBN 0-8103-8375-6.

Travers, Bridget, Ed., *World of Scientific Discovery*. Gale Research Inc.: Detroit, 1994. ISBN 0-8103-8492-2.

Trifonovitch, Kelli Abe, "Hydrogen: Can It Fuel Hawaii's Future?", *Hawaii Business*, July 2004, www.hawaiibusiness.cc

U.S. Department of Energy, Chicago Regional Office, "Distributed Energy Resources in the Midwest: Market Status and Potential," 2002.

U.S. Department of Energy, Office of Energy Efficiency and Renewable Energy, "Hydrogen Fuel," Consumer Energy Information: ERECE Reference Briefs, www.eere.energy.gov/consumerinfo/refbriefs/a109.html, February 2003.

U.S. Department of Energy, Office of Energy Efficiency and Renewable Energy, DOE *Hydrogen Posture Plan*, February 2004, www.eere.energy.gov/hydrogenandfuelcells/pdfs/hydrogen_posture_plan.pdf.

U.S. Department of Energy, Office of Energy Efficiency and Renewable Energy, *National Hydrogen Energy Roadmap*, November 12, 2002, www.eere.energy.gov/hydrogenandfuelcells/pdfs/bk28424.pdf.

U.S. Energy Information Administration, *Annual Energy Outlook 2004 with Projections to 2025*, www.eia.doe.gov/oiaf/aeo.

U.S. Energy Information Administration, *International Energy Outlook 2004*, www.eia.doe.gov/oiaf/ieo.

"U.S. merchant and on-site hydrogen sales to reach $2.7 billion by 2008," *Power Engineering*, October 13, 2003.

Vaitheeswaran, Vijay V., *Power to the People: How the Coming Energy Revolution Will Transform an Industry, Change Our Lives, and Maybe Even Save the Planet*. Farrar, Straus and Giroux: New York, 2003. ISBN 0-374-23675-5.

Valenti, Michael, "Fill'er up—with hydrogen," *Mechanical Engineering*, February 2002, www.memagazine.org.

Vincent, Bill, "Hydrogen and the Law: Safety and Liability," Breakthrough Technologies Institute, Presentation to George Washington University Law School, June 11, 2004.

Wallace, John M., and Christy, John R. , "The Truth About Global Warming," February 4, 2000, www.nas.edu/nrc.

Walters, Russell C., "Fuel Cell Tutorial," Iowa State University, http://erl.cce.iastate.edu.

Wilburn, Richard, "The Development of the Hydrogen Economy and National Energy Policy," www.energypulse.net, February 12, 2004.

Wilson, John R., and Burgh, Griffin, "The Hydrogen Report: An Examination of the Role of Hydrogen In Achieving U.S. Energy Independence," TMG/The Management Group, July 2003, www.tmgtech.com.

Wurster, Reinhold, "Hydrogen Road Vehicles and their Refuelling Infrastructure: Strategic Considerations and The European Situation in Regulations & Standards," L-B-Systemtechnik GmbH, presentation to the H$_2$IT Associazione italiana idrogeno e celle a combustibile, Milan, April 16, 2004.

Zink, John C., "Trend Favors Nuclear-Hydrogen Economy," *Power Engineering*, April 2003, http://pe.pennwellnet.com.

Websites

Air Products and Chemicals, www.airproducts.com

American Council for an Energy-Efficient Economy, www.aceee.org

American Honda Motor Co., Inc., www.honda.com

American Wind Energy Association, www.awea.org

Analytic Power LLC, www.analytic-power.com

Australian Greenhouse office, www.greenhouse.gov.au/ggap

Avalence, LLC, http://avalence.com

Aviation Tomorrow, www.aviationtomorrow.com

Ballard Power Systems, www.ballard.com

BC Hydro, www.bchydro.com

BMW AG, www.bmw.com/bmwe/pulse/enterprise/cleanenergy3

Boeing, www.boeing.com

BP plc, www.bp.com

Bureau of Transportation Statistics, www.bts.gov

Business for Social Responsibility (U.S.), www.bsr.org

California Air Resources Board, www.arb.ca.gov

California Fuel Cell Partnership, www.cafcp.org, www.drivingthefuture.org

California Hydrogen Business Council, www.ch2bc.org/

California Hydrogen Highway, www.hydrogenhighway.ca.gov

Canadian Transportation Fuel Cell Alliance, www.ctfca.nrcan.gc.ca

Capstone Turbine Corp., www.capstoneturbine.com

Carbon Sequestration Leadership Forum, www.cslforum.org

Casio, www.casio.com

Catalytica Energy Systems, www.catalyticaenergy.com

ChevronTexaco, www.chevrontexaco.com

DaimlerChrysler, www.daimlerchrysler.com

Database of State Incentives for Renewable Energy (DSIRE), www.dsireusa.org.

Dow Chemical Co., www.dow.com

DTE Energy, www.dteenergy.com

Ebara Corp., www.ebara.co.jp

Electric Power Research Institute, www.epri.com

European Commission, CUTE project, http://europa.eu.int/comm/energy_transport/en/prog_cut_en.html#cute

European Commission, Energy Research Program, http://europa.eu.int

European Environment Agency, www.eea.eu.int

Ford Motor Co., www.ford.com; www.fordvehicles.environment

Fraunhofer Institute, www.ise.fhg.de

FreedomCAR, www.eere.energy.gov/vehiclesandfuels/program_areas/freedomcar

Fuel Cell Bus Club, www.fuel-cell-bus-club.com

Fuel Cell Today, www.fuelcelltoday.com

Fuel Cells 2000—The Online Fuel Cell Information Resources, www.fuelcells.org

FuelCell Energy, Inc., www.eerc.com

General Motors Corp., www.gm.com

Harvest Energy Technology, Inc., www.harvest-technology.com

Hawaii Natural Energy Institute (HNEI), www.hnei.hawaii.edu

Heliocentris Energy Systems, www.heliocentris.com

HERA Hydrogen Storage Systems Inc., www.herahydrogen.com

Honda Motor Co., Ltd., http://world.honda.com

Hydrogen Appliances, www.hydrogenappliances.com

Hydrogen Burner Technology, Inc., www.hbti.net

HydrogenSource, www.hydrogensource.com

HyNet—The European Hydrogen Energy Thematic Network, www.hynet.info

Iceland New Energy, Ltd., www.ectos.is

IdaTech, www.idatech.com

Illinois Coalition, www.ilcoalition.org

Ingersoll Rand Co., www.ingersoll-rand.com

InnovaTek, Inc., www.tekkie.com

Iwatani International Corporation of America, www.iwatani.com

Japan Hydrogen & Fuel Cell Demonstration Project, www.jhfc.jp/e

Johnson Matthey Gas Purification Technology Group, www.hydrogentechnology.com

Johnson Matthey, www.jmusa.com

Lawrence Livermore National Laboratory, www.llnl.gov

Linde AG, www.linde.com

Masterflex, www.masterflex.de

Methanol Institute (U.S.), www.methanol.org

Millennium Cell, Inc., www.millenniumcell.com

Ministry of Economy, Trade and Industry, Japan, www.meti.go.jp/english

Mitsubishi Motors, www.mitsucars.com

MTU CFC Solutions GmbH, www.mtu-friedrichshafen.com/cfc/en/cfcs/cfcs.htm

Musashi Institute of Technology, www.musashi-tech.ac.jp

National Hydrogen Association, www.hydrogenus.org

Natural Resources Canada, www.nrcan-rncan.gc.ca

NEG Micon A/S, www.neg-micon.com

NextEnergy, www.nextenergy.org

Nissan, www.nissan-global.com

Norsk Hydro, www.hydro.com

Nu Element Inc., www.nuelement.com

Nuvera Fuel Cells, www.nuvera.com

Partnership for Advancing the Transition to Hydrogen, www.hpath.org

Plug Power, www.plugpower.com

Proton Energy Systems, www.protonenergy.com

Quantum Technologies, Inc., www.qtww.com

Rocky Mountain Institute, www.rmi.org

SHEC labs, Solar Hydrogen Energy Corp., www.shec-labs.com

Shell, www.shell.com

Shell Hydrogen, www.shell.com/hydrogen

Shell Solar, www.shell.com/solar

Siemens/Westinghouse, www.siemens.com

SOFCo-EFS, www.sofco-efs.com

South Coast Air Quality Management District, www.aqmd.gov

Stuart Energy Systems Corp., www.stuartenergy.com

Teledyne Energy Systems, Inc., www.teledynees.com

Texaco Ovonic, www.txohydrogen.com

The Japan Steel Works, Ltd., www.jsw.co.jp/en

Toshiba, www.toshiba.com

Toyota, www.toyota.com

U.S. Department of Energy, Office of Energy Efficiency and Renewable
    Energy, Hydrogen, Fuel Cells, and Infrastructure Technologies Program,
    www.eere.energy.gov/hydrogenandfuelcells

U.S. Energy Information Administration, Status of State Electric Industry Restructuring Activity, www.eia.doe.gov/cneaf/electricity/chg_str/regmap.html

U.S. Energy Information Administration, www.eia.doe.gov

United Nations Framework Convention on Climate Change, http://unfccc.int

United Technologies Corp., UTC Power Division, UTC Fuel Cells, www.utcfuelcells.com

Wah Chang, www.alleghenytechnologies.com/WahChang

Wellman CJB Ltd., www.wellmangroup.co.uk/htm/CJB.htm

Wired, www.wired.com

World Energy Network (WE-NET), www.enaa.or.jp/WE-NET

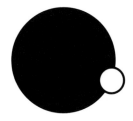

Index

D

F

H

J–K

L

M

N

O

P

U

V

W–Y

Z